Biotechnology Monographs

Volume 5

Editors

S. Aiba · L.T. Fan · A. Fiechter · J. Klein · K. Schügerl

C. R. Phillips   Y. C. Poon

Immobilization of Cells

With 24 Figures

Springer-Verlag
Berlin Heidelberg New York
London Paris Tokyo

Professor Dr. Colin R. Phillips
Dept. of Chemical Engineering
and Applied Chemistry
University of Toronto, Ontario
Canada M5S 1A4

Dr. Yiu Cheong Poon
Dept. of Chemical Engineering
and Applied Chemistry
University of Toronto, Ontario
Canada M5S 1A4

Library of Congress Cataloging-in-Publication Data.
Phillips, C. R. (Colin Rex), Immobilization of cells. (Biotechnology monographs; v. 5)
Includes bibliographies and indexes. 1. Immobilized cells. 2. Immobilized cells – Indus-
trial applications. I. Poon, Y. C. (Yiu Cheong). II. Title. III. Series. TP248.25.I55P48
1988   660'.6   88-3066
ISBN-13: 978-3-642-73251-5   e-ISBN-13: 978-3-642-73249-2
DOI: 10.1007/ 978-3-642-73249-2

© Springer-Verlag Berlin Heidelberg 1988
Softcover reprint of the hardcover 1st edition 1988

Typesetting: Brühlsche Universitätsdruckerei, Giessen
2152/3145-543210 Printed on acid-free paper

Preface

Growth of immobilized cells can be viewed as an alternative to growth of free cells in many instances. In others, immobilization confers a precision of control over the process not possible in free growth. Immobilization of cells can sometimes be considered to be a lower cost alternative to immobilization of enzymes. In this volume, immobilization procedures based on mechanical means and bonding of various types are examined, with detailed application examples. These applications include microorganisms, plant and animal cells, sub-cellular organelles and multiple enzyme systems. Particular attention is devoted to enzyme properties in immobilized cells and the properties of the carrier. The volume should provide the reader with a comprehensive overview of the subject, together with copious references. As well as serving as a research monograph, it could be used to provide reference material for a graduate course.

Special thanks are due Mrs. JENNIFER KERBY for her dedicated work in the preparation of the manuscript, and IT-CHIN HSIEH for bibliographical assistance.

Toronto, July 1988

COLIN R. PHILLIPS
YIU C. POON

Table of Contents

1 Introduction

That living microorganisms can attach to each other and to solid surfaces in the form of films is well known [1]. Such attachment occurs in dental plaque, algal and fungal slimes, bacterial film on soil particles [2], on plant or animal tissues such as intestine and rumen [3] and in chicken crop epithelia [4]. The same effect is also observed in the tendency of many bacteria to assume chain structure. Cell adherence, a phenomenon of great importance in monolayer cell culture because it allows easy manipulation of growing cells, has recently been reviewed [5, 6].

Bacterial films, a consequence of bacterial growth and adherence, are so ubiquitous in nature that it may be safe to conclude that wherever there is a solid substrate there is likely to be a bacterial film involved. Examples of early utilization of bacterial films include the production of vinegar by the trickling method, the history of which is reviewed by CONNER and ALLGEIER [7], and the leaching of mineral ores by sulphur oxidizing bacteria [8].

From as early as 1670, vinegar making processes were known in which packings of various plant materials such as grape vine, rape, corncobs, wood, as well as inorganic materials such as pumice, ceramics and charcoal, were used as support for the slime film of the *Acetobacter* microorganisms. Substrate, air, and nutrient solutions were trickled through layers of those packings and converted to the product. This process is inherently adaptable to continuous flow fermentation, and packings of beech wood shavings have been in continuous use in some plants for 50 years [7].

In the field of hydrometallurgy, heap leaching of copper ores [8] was practised in Spain in 1725. There is a Chinese record [9] in the twelfth century of extracting copper from copper sulphate dissolved in natural waters, probably as a result of subterranean leaching, although there was no conscious effort to utilize bacterial action in leaching, nor recognition of its presence. The original dissolution is due to the action of various strains of sulphur oxidizing bacteria, such as those in the genus *Thiobacillus,* on sulphide minerals to produce soluble sulphates. Members in this genus are diverse, including aerobic, anaerobic and faculative anaerobic bacteria growing under different conditions in solution as well as on solid surfaces. On the other hand, sulphate reducing bacteria are generally considered to be responsible for the formation and deposition of sulphide ores in nature.

Processes in which localized bacterial films are active include the aging of meat to improve its texture and flavour, the production of Rochefort cheese and similar solid fermented food, composting, the treatment of waste, whether intentional or not, by trickling it over bacterial slime, the fixation of nitrogen by bacteria attached to roots of leguminous plants, and decomposition and rotting in nature.

Although bacterial films occur often in nature and have found limited application in vinegar production, ore leaching and the preparation of some fermented food, they were not widely utilized in industrial fermentation reactions until the 1970's, when the demand for viable alternatives to enzyme fermentation became increasingly urgent. The theoretical and morphological basis for cell attachment behaviour, however, has commanded much more interest, more with respect to fundamental understanding than to application. Current theories of cell adhesion have been extensively reviewed [5, 6].

The technique of cell immobilization is an outgrowth of enzyme fermentation. Because of their specific catalytic activity and their high performance under mild physiological conditions, enzymes have become increasingly important in fermentation. Enzymes can be used in the free form or in an insoluble form in aqueous systems. In these two forms, enzymes can also be bound to a solid carrier which, in many instances, results in increased operational stability, enhanced activity, and the possibility of a continuous process in which better control of substrates and product flow can be maintained. Nevertheless, enzyme processes suffer from certain disadvantages, the most serious of which are the cost of pure enzymes, the difficulty of recycling them by extraction, and problems of product contamination by leaking, in the case of some immobilized enzymes. To circumvent these disadvantages, the more readily available microbial enzymes, together with the cells containing them, are bound to carriers by various methods, often with remarkable improvements in enzyme activity and half-life.

Although the presence of various reagents and the procedures of drying, freezing, crosslinking, etc. inherent in immobilization may kill the cells, their enzyme activity may not be impaired. In fact, in fermentations requiring single intracellular enzymes, immobilization processes usually increase the permeability of the cell walls, and allow nutrients, substrates and products to diffuse more readily through the cell membranes and thus facilitate the fermentation. The dead cells in this sense act also as a carrier. More complex immobilizations may involve living cells containing multiple enzymes and subcellular components having particular functions in a fermentation reaction.

Early Work on Enzyme Immobilization
The history of cell immobilization is closely associated with that of enzymes [10–12]. Empirical studies of enzyme immobilization, mainly by adsorption onto inert solids, follow a pattern of experimental work in the early 1900's leading to the formulation of the Michaelis-Menten hypothesis of enzyme catalysis (Table 1.1). Thus as early as 1908, MICHAELIS and EHRENREICH studied the adsorption of a range of enzymes on various solid adsorbents as a function of pH, and found that adsorption is pH dependent [13]. Adsorption was found to be irreversible in the case of peptone on animal charcoal [14] and also dependent on the ionic character of the adsorbent [15]. From these early observations followed a series of studies on the electrophoresis of enzymes [16–18]. The dependence of adsorption on the concentration of the enzyme in solution was studied by JACOBY in 1916 [19]. The 1920's heralded a period of application of the principles of enzyme adsorption, mainly in the area of enzyme purification. The status of enzyme purification by adsorption techniques was reviewed by FABRE in 1923 [20].

2

Table 1.1. Early examples in adsorption of biological materials enzymes and polypeptides

Materials adsorbed	Carrier	Description	Ref.
Diastase, invertase, pepsin, trypsin, rennin	Kaolin, talc, animal charcoal, Al_2O_3, Fe_2O_3	Extent of adsorption	[13]
Peptone, albumose	Animal charcoal	Reversibility of adsorption	[14]
Zymose		Adsorption, affinity	[15]
Invertin		Electrophoresis	[16]
Pepsin		Electrophoresis	[17]
Malt diastase	Kaolin	Electrophoresis	[18]
Pepsin	Al_2O_3		[63]
Albumin-peptone	Al_2O_3		[64]
Invertase	Animal charcoal, aluminum hydroxide		[65]
Urease	Fibrin flakes	Activity and contraction of adsorbed enzyme	[19]
Proteins, enzymes, toxins, and sera	Aluminum hydroxide	Purification studies	[21]
Sucrase, amylase	Aluminum hydroxide		[22]
Sucrase	Kaolin, aluminum hydroxide	Absorption affinity	[23]
Amylase	Aluminum hydroxide	Function of ethanol as co-absorbents	[24]
Sucrase	Aluminum hydroxide	Adsorption isotherms	[66]
Invertin		Purification by adsorption	[67]
Pepsin	Animal charcoal	Selective elution	[25]
Diastase	Animal charcoal	Temperature effects, irreversibility	[26]
Invertin	Kaolin, aluminum hydroxide		[23]
Amylase	Alumina gel	Enzyme purification	[29]
Papain	Aluminum hydroxide	Procedures of purification	[30]
Serum enzyme	Various adsorbents	Deactivating power of adsorbents	[31]
Pepsin, rennin, catalase, peroxidase	Cellulose	Strength of adsorption	[33]

Aluminum hydroxide was generally selected as the adsorbent in the purification of various biological substances such as proteins, toxins, enzymes, and antitoxins [21]. Adsorption studies of various enzymes on $Al(OH)_3$ and animal bone charcoal were undertaken to examine the physical and chemical behaviour of adsorption, for example, the adsorption isotherm [22, 23], effects of pH, solvent and inhibitors [24] and the reversibility of adsorption [25, 26]. Such fundamental studies provide the basis for the use of adsorbents in the purification of enzymes, still an important aspect of enzyme technology.

Early attempts at enzyme purification are reviewed by WILLSTÄTTER [27] in 1926 and later in 1932 [28]. The main subjects of investigations from the 1920's to the 1950's were adsorption efficiency as a function of pH [29], inactivation by adsorbents [30, 31], inhibition and activation [32], and selectivity of adsorption [33]. These topics were reviewed by ZITTLE [34] in 1953.

Table 1.2. Early examples in adsorption of biological materials

Materials adsorbed	Carrier	Description	Ref.
Bacteria	Animal and vegetable charcoal, Fuller's earth	Differential staining	[44]
Bacteria and Agglutinins	Animal charcoal, barium sulphate, aluminum hydroxide	Bacteria agglutination	[45]
Bacteria	Soil	Adsorption capacity	[52]
Bacteria	Soil		[51]
Bacteria		Adsorption of iodine on bacteria	[46]
Bacteria		Behaviour of gram-positive and gram-negative bacteria on iodine adsorption	[47]
Bacteria		Disinfection by iodine	[48]
Bacillus caryocyaneus		Adsorption of Bordeaux B	[50]
Bacillus caryocyaneus		Phenolized gentian violet staining	[68]
Bacteria, moulds, yeasts	Industrial clay	Identification of micro-organisms	[69]
Azotobacter chroöcoccum	Loams	Influence of adsorbed ionic species	[70]
Various bacteria	Kaolin, $CaCO_3$, $Al(OH)_3$, $BaSO_4$	Selective adsorption	[55]
Azotobacter	Dispersed sand	Study of parameters in adsorption	[56]
Bacteria	Pectin, white clay, active charcoal, lignin	Bacterial growth	[59]
Bacteria	Filter paper	Study of antibiotic sensitivity	[60]
Bacteria	Soil	pH dependence of adsorption	[71]
Bacteria	Soil	Desorption of bacteria from soil	[72]

The period from the 1930's to the 1960's is characterized by intensive work on particular applications, methods of immobilization, and the use of synthetic and natural polymers (Table 1.2). In the 1930's, WHITE in the US and GAUTHERET in France developed completely synthetic media for plant cells in tissue culture. Industrial and research applications of enzymes immobilized on polymer supports were extensively investigated, notably in the field of immunology. MANECKE developed a number of synthetic resins as specific adsorbents for serum proteins [35–37].

By 1953, some enzymes were successfully immobilized on polymer solids, and patents were granted for a number of techniques developed by KATCHALSKY throughout the 60's [38–40]. At this time, several chemical reagents were developed for enzyme immobilization, in addition to the organic and inorganic solid adsorbents. The new immobilization reagents formed covalent bonds between enzymes and the solid support, and include diazotized polyaminostyrene, resin containing carboxylic acid chloride functional groups and diazotized aminobenzylcellulose. A review of these chemical immobilization methods has been made by BRANDENBERGER [41].

In 1957, BRANDENBERGER [42] reviewed all the current techniques, five of which he developed, for binding physiologically active proteins and enzymes onto solid supports. The amount of carbon dioxide liberated from the hydrolysis of the unreacted isocyanate groups after the immobilization of a protein on a polyisocyanate carrier was used as a criterion for covalent bonding in immobilization.

In the 1960's increasingly frequent application of enzymes was made to fermentation. Since that time, research work on immobilization has multiplied enormously, much of it with an industrial focus. The two processes, enzyme immobilization and cell immobilization, have a common mode of action, that is, heterogeneous enzyme catalysis; it should be noted that lysis of microbial cells after their immobilization was one of the early methods of immobilizing enzymes. However, numerous new procedures have been developed for immobilization of cells so that it now stands by itself as a separate process, often complementary to that of enzyme immobilization. The first successful industrial production using an immobilized enzyme was based on aminoacyclase, and was developed by CHIBATA et al. [43] in 1969. Racemic mixtures of amino acids were optically resolved through selective enzymatic action to yield the D- and L-optical isomers. Since then several other industrial scale production fermentation processes using immobilized biological substances have been developed.

Early Work on Cell Immobilization
Study of the attachment behaviour of microorganisms has always lagged behind that of enzymes. The lag might be attributed to the greater inherent complexity of microbial cells relative to enzymes. As discrete chemical compounds, enzymes are easier to purify, characterize and assay. Initially, the only method of attachment of microbial cells to a solid was by adsorption, which remains the most direct among the modern techniques, although it is not necessarily the most effective or easiest.

Early study of bacterial adsorption was limited to choice of a solid support and suitable conditions, study in soil systems, study of activity of the adsorbed microbes, and later, after development of the gram positive and gram negative classifications of microorganisms, study of the selectivity of adsorption of bacterial cell surfaces. The adsorbing powders of animal and vegetable charcoal and Fuller's earth were found to be effective in 1918 [44]. The dependence of adsorption on particle size and the "surface development" of the solid support was investigated using basic methylene blue and acid dyestuffs. BLEYER [45] in 1922 studied the adsorption behaviour of the agglutinins on suspensions and colloids and found animal charcoal and fine precipitates such as $BaSO_4$ to be the most effective. The dispersed agglutinin was shown to agglutinate homologous bacteria preferentially over absorbents in colloidal state, even though the adsorption of agglutinin on the colloid was found to be stable with heating and in 0.01 N NaOH.

Later in 1932, a series of investigations was conducted on the adsorption of iodine [46] on gram positive and gram negative bacteria [47] in relation to the action of disinfection [48]. DIETZEL et al. surveyed the then-current theoretical and practical application of the bactericidal properties of various absorbent preparations [49]. The effect of pH on the adsorption of the dye Bordeaux B on the cells of *Bacillus caryocyaneces* was investigated by LASSEUR and DUPAIX-LASSEUR [50].

Adsorption studies of bacteria on soils were undertaken by KHUDIAKOV [51] to determine the flocculating behaviour of the bacteria, the activity of the adsorbed cells and the capacity [52]. In 1936, GLICK [53] found that of all the microorganisms found on samples of industrial clay only the aerobic or facultative bacteria showed an increase in viability after 3 months storage at the optimum temperature of 30 °C. PEELE [54] used CO_2 evolution as a measure of activity of bacteria adsorbed on soils and postulated that adsorption was due to electrical attraction. Other inorganic solid adsorbents investigated include kaolin, $CaCO_3$, $Al(OH)_3$ and $BaSO_4$ [55]. In 1950, the physical parameters for the adsorption of *Azotobacter* on sand were found to include the electrokinetic potential which is a function of pH and which could be used as a criterion for reversible or irreversible adsorption (adhesion) [56]. This work is the forerunner of the study of bacterial adhesion by the zeta potential, a topic extensively reviewed by GERSON and ZAJIC [6], LIPS and JESSUP [57], and ROGERS [58].

Organic adsorbents used in this period include pectin [59] and filter paper impregnated with methyl cellulose [60].

Evaluation of Whole Cell Immobilization as a Process
Evaluation of cell immobilization as a process relative to alternative processes is highly complex, and must take into account a great number of factors, the foremost of which are economic factors, environmental factors, and, in the case of production of food stuff, taste and aesthetics. Economic factors include cost of starting materials, energy and equipment, and required technical skill. Thus comparison of the advantages and disadvantages of these processes can only be made in a very general way, with the assumption that all other factors are constant.

Relative to conventional chemical processes, the most basic difference has to do with the nature of the process, that is, chemical synthesis vs. biological (enzyme) fermentation. In general, enzymatic reactions are much more efficient and specific as to identity and stereochemistry of products and conditions of reaction. Organic synthesis reactions on the other hand, although more various in nature, often require harsh conditions in temperature and pressure, use of organic solvents, extremes of pH and less selective but very reactive reagents. Transformations in organic reactions are particularly sensitive to steric crowding which can effectively preclude the possibility of reaction, no matter how much driving force is used.

Enzyme catalysis has the additional advantages of adaptability to low grade substrates such as recycled waste material or surplus material from other industries, and the availability of a number of metabolic pathways which may lead to different products by careful control of conditions and substrates. Biological catalysis reactions generally cause less pollution problems. In contrast to these advantages, biological catalysis reactions suffer from several disadvantages, the

6

most serious of which are the instability of enzymes and microbial cells and the need to carefully maintain external conditions and to supply the requisite cofactors, all of which add to the expense of the process. Enzymes in the purified form are usually expensive. Further, because of their high reactivity, biological catalysis by microbial cells often leads to byproducts which add to the problem of purification and extraction. In the free form, the enzyme is difficult to recycle. Neither enzymes nor cells are strong mechanically and their mechanical properties often impose constraints on the design of fermentation equipment. Other constraints include restrictions in the form of physical conditions such as pH, temperature, pressure, the nature of the solvent system, and the presence of inhibitors.

Among biocatalyzed processes, the comparison between whole cells (or organelles) and enzymes (or coenzymes), in their free or immobilized forms, is complex. The generally recognized situation is that immobilized enzyme processes are outgrowths of free cell fermentations and immobilized cell processes are outgrowths of immobilized enzyme processes. Thus immobilized systems are looked upon as alternatives to the established free cell fermentations only when they offer distinct advantages over existing processes. Valid comparisons may be possible between processes involving immobilized cells and free cells, and between processes involving immobilized cells and immobilized enzymes.

Immobilized Cells versus Free Cells

If the comparison is limited to preparative fermentation processes, as distinct from such processes as waste disposal and mineral leaching, where the cost of the pure enzyme or the microorganisms is often of secondary importance, the advantages of immobilized cells are mainly with respect to general productivity and operational flexibility.

Firstly, the productivity of immobilized cells is generally as high as, if not higher than the corresponding free cell fermentations. This productivity can be explained by the fact that the microenvironments offered by the carrier are more stabilizing to the organism or its enzymes, which generally show optimal activity only under very narrowly prescribed physical conditions. In by far the great majority of the reported examples of cell immobilization, regardless of whether dead or viable cells are involved, enzyme activity is enhanced, as indicated by a longer half life. For living cells, the rate of growth, indicated by the generation time, is also increased by immobilization. Consequently both the operational stability of the immobilized organisms and the productivity are improved. Thus, taken as a whole, enzymes in immobilized cells, whether viable or not, retain their activity better than in free cells.

Another advantage of immobilized cell systems is that they can be described readily by well developed theoretical and hydrodynamical treatments of heterogeneous catalysis systems, particularly if the system utilizes solid carriers in the form of uniform spherical particles.

With respect to operational flexibility, comparison is not as straightforward. In general, immobilized cells make possible continuous fermentations which do not work very well with free cells. Because of the possibility of higher cell loadings, reaction rates may be higher. Cell immobilization also enables higher dilution without culture washout, and greatly facilitates recycling or reuse of microorganisms. A greater degree of control is possible, mainly in the areas of selectivity in maintaining the stage of cell activity at maximum productivity, in reactivation or regeneration of enzyme activity, in selective isolation of inhibitory reaction intermediates or products, in enhancement of productivity by co-immobilization of metabolically complementary microorganisms involving the transfer of a wide range of substances between cells, in product separation and, finally, in waste disposal. Nevertheless, cell immobilization systems do suffer from a number of problems. The initial expense for such a system is usually high and the process usually requires a large reactor. The mechanical properties of the system – microbial cell bound to solid carrier – are more complex than those of free cells and have to be taken into account in order to provide a continuous, recycling process and effective agitation and filtration. System designs must allow for the increased diffusion barrier through the cell and the carrier for substrates, products, and cofactors, so that cells bound to a carrier generally need permeabilizing treatment, especially when high molecular weight substrates or products are encountered.

Reagents such as monomers, crosslinking reagents, and radical initiators used in the polymerization of synthetic organic carriers are often cytotoxic and the organic solvents used in these reactions can lyse microorganisms and denature the enzymes contained in them.

Immobilized Cells versus Immobilized Enzymes

With respect to preparative fermentations, the important determining factors in comparing immobilized cell processes with immobilized enzyme processes are the cost and the nature of the reaction process. It is generally recognized that the use of an enzyme is more costly in terms of initial cost, extraction, purification, recycling, reactivation, decontamination and final disposal. In many processes, these costs will determine the choice between the two alternatives. The nature of the reaction process is important in terms of whether the reaction is single enzyme (intracellular) catalyzed or multiple enzyme catalyzed, involving the whole metabolic system of the cell, and requiring many sequential enzymes, coenzymes and cofactors. Multiple enzyme systems immobilized on a carrier are generally used for investigative studies rather than for production. On the other hand, whole cells that produce a required metabolite have been used co-immobilized with enzymes, although mostly in experimental studies. In general, a natural cellular environment contributes to the stability and activity of the enzyme and the cell membrane offers some protection against detachment of the enzyme. Enzymes so contained can assume their natural structural conformation, both with respect to the carrier and to the substrate. With whole cells there is the further advantage of controlling the growth stage of the cells to maximize product conversion. Balanced against these advantages, immobilized whole cells have the disadvantage of an increased diffusion barrier, especially for high molecular weight substrates and products. Because of the more complex biochemical make-up of the whole cells, side reactions are more likely, and sterile conditions may be required. Some methods of immobilization, for example, covalent attachment, or crosslinking, invariably kill the cells and thus preclude their use in multienzyme fermentations. At present, in industrial production, the two immobilization systems – microbial whole cells and single enzymes – are at about the same stage of development [61, 62].

Scope of Present Book
The literature on immobilized whole cells has burgeoned since the early 1970's. In addition to a great number of review articles, several books have been written. In the present work, the literature on immobilized cells, organelles and other sub-cellular materials is reviewed to provide a survey complementary to presently available publications. Topics such as affinity chromatography, dialysis culture, cell culture, and immunology, are not covered because of the limitations of the present work and the availability of extensive existing literature.

References

1. Fletcher M, Floodgate GD (1976) In: Fuller R, Lovelock DW (eds) Microbial ultrastructure. Academic Press, New York, p 101
2. Burns RG (1979) Microorganisms and soil surfaces. In: Ellwood DC, Melling J, Rutter P (eds) Adhesion of microorganisms to surfaces. Academic Press, New York, p 109
3. Knapp JS, Howell JA (1980) Top Enzyme Ferment Biotechnol 4:85
4. Brooker BE, Fuller R (1976) In: Fuller R, Lovelook DW (eds) Microbial ultrastructure. Academic Press, New York, p 85
5. Ellwood DC, Melling J, Rutter P (eds) (1979) Adhesion of microorganisms to surfaces, society for general microbiology. Academic Press, New York
6. Gerson DF, Zajic IE (1979) In: Venkatsubramanian K (ed) Immobilized microbial cells. ACS Symposium Series 106. Am Chem Soc, Washington DC
7. Conner HA, Allgeier RJ (1976) Adv Appl Microbiol 20:81
8. Karaviko GI, Kuznetsov SI, Golonizik AI (1977) The bacterial leaching of metals from ores. Technology Limited, England
9. Shen Kuo (1961) Meng, Hsi Pi Tang Chiao Cheng, vol 2. Hsi Chia Book, Taiwan, p 792
10. Chibata I, Tosa T (1983) Appl Biochem Bioeng 4:1
11. Dunnill P (1980) Philos Trans R Soc London B 290:409
12. Jack TR, Zajic JE (1977) Adv Biochem Eng 5:125
13. Michaelis L, Ehrenreich M (1908) Biochem Z 10:283
14. Michaelis L, Rona P (1909) Biochem Z 15:196
15. Michaelis L, Rona P (1909) Biochem Z 15:217
16. Michaelis L (1909) Biochem Z 16:81
17. Michaelis L (1909) Biochem Z 16:486
18. Michaelis L (1909) Biochem Z 17:231
19. Jacoby M (1916) Biochem Z 74:93
20. Fabre R (1923) Bull Soc Chim Biol 5:432
21. Rakusin MA (1922) Z Immunität 34:155
22. Kraut H, Wenzel E (1924) Z Physiol Chem 133:1
23. Kraut H, Wenzel E (1925) Z Physiol Chem 142:71
24. Willstätter R, Waldschmidt-Leitz E, Hesse ARF (1925) Z Physiol Chem 142:14
25. Kikawa K (1926) J Biochem (Japan) 6:275
26. Unna Z (1926) Biochem Z 172:392
27. Willstätter R (1926) Naturwissenschaften 14:937
28. Willstätter R (1932) Naturwissenschaften 20:624
29. Sherman HC, Caldwell ML, Adams M (1926) J Am Chem Soc 48:2947
30. Kraut H, Bauer E (1927) Z Physiol Chem 164:10
31. Dormal J (1927) Compt Rend Sci Biol 97:898
32. Dyckerhoff H, Miehler H, Tadsen V (1934) Biochem Z 268:17
33. Tauber H (1936) J Biol Chem 113:753

34. Zittle CA (1953) Adv Enzymol 14:319
35. Manecke G, Gillert KE (1955) Naturwissenschaften 42:212
36. Manecke G, Singer S, Gillert KE (1958) Naturwissenschaften 45:440
37. Manecke G, Singer S, Gillert KE (1960) Naturwissenschaften 47:63
38. Katchalsky E, Bar-Eli A (1963) Br 916, 931, Jan 30
39. Katchalsky E, Bar-Eli A (1962) Israeli 14,448, Oct 25
40. Katchalsky E, Bar-Eli A (1960) Israeli 13,950, Appln. May 27
41. Brandenberger H (1955) Congr Intern Biochim, Resumés Communs, 3è Congr, Brussels, p 29
42. Brandenberger H (1956) Rev fermentations et inds aliment 11:237
43. Chibata I (1979) Food Proc Eng 2:1
44. Bechhold H (1918) Kolloid Z 23:35
45. Bleyer L (1922) Z Immunitäts Abt I Orig 33:478
46. Habs H (1932) Z Hyg Infektionskrankh 113:239
47. Habs H (1932) Z Hyg Infektionskrankh 114:1
48. Habs H (1932) Z Hyg Infektionskrankh 114:358
49. Dietzel R, Schlemmer F, Hamann V (1932) Apoth Ztg 47:244, 261, 283
50. Lasseur Ph, Dupaix-Lasseur A (1934) Trav lab microb faculté pharm Nancy 7:123
51. Khudiakov NN (1926) Pochvovedenie 21:46
52. Chudiakow NN (1926) Centr Bakt Parasitenk II Abt 68:345
53. Glick DP (1936) J Am Ceram Soc 19:169
54. Peele TC (1936) Agr Expt Sta Memoir 197:3
55. Gunnison JB, Marshall MS (1937) J Bact 33:401
56. Tschapek W, Garbosky AJ (1950) Trans 4th Intern Congr Soil Sci Amsterdam 3:102
57. Lips A, Jessup NE (1979) In: Ellwood DC, Melling J, Rutter P (eds) Adhesion of microorganisms to surfaces. Society general microbiology. Academic Press, New York, p 5
58. Rogers HJ (1979) In: Ellwood DC, Melling J, Rutter P (eds) Adhesion of microorganisms to surfaces. Society general microbiology. Academic Press, New York, p 29
59. Debusmann M (1950) Monatsschr Kinderheilkd 98:336
60. Ryan WL (1961) US 2,998,353, Aug 29
61. Cheetham PS (1980) Topics Enzyme Ferment Biotechnol 4:189
62. Kolot FB (1980) Process Biochem 15:2
63. Rakuzin MA, Brandon EM (1915) J Russ Phys Chem Soc 47:1055
64. Rakuzin MA, Brandon EM (1915) J Russ Phys Chem Soc 47:1057
65. Nelson JM, Griffin EG (1916) J Am Chem Soc 38:1109
66. Willstätter R, Wenzel E (1925) Z Physiol Chem 142:71
67. Willstätter R, Wenzel E (1925) Z Physiol Chem 142:257
68. Lasseur P, Benoit M (1934) Trav Lab Microbiol Faculté Pharm Nancy 7:129
69. Dudley I, Glick P (1936) J Am Ceram Soc 19:169
70. Peele TC (1936) N.Y. (Cornell) Agr Expt Sta, Memoir 197:3
71. Samsevich AS (1939) Chemisation Socialistic Agr (USSR) No 12:37; (1940) Khim Referat Zhur No 5:61
72. Zvyagintsev DG (1962) Pochvovedenie, No 2:19

2 Methods of Cell Immobilization

Methods of cell immobilization roughly parallel those of enzyme immobilization and can best be classified by the nature of the mode of attachment, that is, as mechanical, chemical or ionic. In mechanical immobilization, the cells are localized by means of physical barriers. In chemical immobilization, covalent bonds are formed among cells or to a solid phase. In ionic immobilization, electrostatic, van der Waal's or London forces of attraction are present. Cells can also attach themselves to solid supports in the course of natural growth, using a combination of these means. This classification is obviously not clear-cut but does serve the purpose of organizing the diverse methods of immobilization available. In Table 2.1, examples of cell immobilization are classified by mode of attachment.

2.1 Mechanical Immobilization

2.1.1 Mycelial Pellet and Mat

The mycelium of a fungus consists of tubular filaments (hyphae) on the top of which are the spores. Typically the hyphae of a fungus are about 50 microns in length and under active growth will branch out, interweave and fuse with one another to form a net. Occasionally such mycelial mats cause problems in fermentations because overcrowding of mycelial growth will cause inactivation in a fermentation using filamentous cells, such as *Rhizopus nigricans* [1] in the hydroxylation of the steroid progesterone, probably due to reduction of the surface area of contact between the medium and the fungus so that the nutrients and oxygen become less accessible.

The growth and citric acid production of *Saccharomycopsis lipolytica* in a trickle flow fermentor containing wood chips as the solid carrier have been studied [2, 3]. Kinetic data indicate that acid production, which starts after a linear growth phase, follows a constant specific rate for 80 h. Studies with simple models confirm that limitations in oxygen diffusion as well as metabolic modification in the immobilized cell are responsible for the 30% reduction in growth and citric acid production compared with free cell fermentations. However, in at least one instance, namely, the production of citric acid by the organism *Aspergillus niger*, where the supply of oxygen has to be controlled, the formation of mycelial pellets actually ensures that a limited supply of oxygen is maintained for optimum citric acid production [4]. Cells of *Streptomyces* containing glucose isomerase have been heat-treated, mixed with an 8% citrate solution at pH 6.0 for 1.5 h, then air-dried.

Table 2.1. Examples of cell immobilizations

(I) Mechanical Attachment

(a) on mycelium mass

Cell	Microbial enzyme	Solid support	Substrate	End Result	Ref.
Saccharomycopsis lipolytica		wood chips	glucose		3
Morterella vinacea	α-galactosidase		p-nitrophenyl(α-D-galactopyranoside)(PNPG)	hydrolysis	7
Absidia griseola var izuchii				α-galactosidase	8
Streptomyces	glucose isomerase	heat fixing	glucose		157
Rhizopus nigricans	hydroxylase	agar alginate gels	progesterone	11 α-hydroxylation	1
Aspergillus ochraceus		mycelium pellets	N-acetyl-DL-methione	ℓ-methionite	9
Escherichia coli, Salmonella typhimurium, Bacillus sp.			18-crown-6		11
Escherichia coli.		filamentous cells	18-crown-6		12
Saccharomycopsis lipolytica		wood chips	glucose	citric acid	2

(I) Mechanical Attachment

(b) by encapsulation

Cell	Microbial enzyme	Solid support	Substrate	End Result	Ref.
Saccharomyces cerevisiae		ethyl cellulose	AMP + glucose + P_i	ATP	17
Micrococcus denitrificans ATCC 21909		oil droplet	NO_3^-, nitrite	Reduction	18
Comamonas sp.		cellulose acetate microcapsules	3-acetoxymethyl-7-(4-carboxy-butane amide)-3-cephan-4-carboxylic acid	7ACA	158
Streptomyces		ethylcellulose capsules	glucose	fructose	159
Streptomyces sp.		polyester capsules	glucose	fructose	160
Streptomyces sp.	glucose isomerase	polyester capsules	xylose		161
Trichoderma E-58 mycelium	β-glucosidase	encapsulated calcium alginate	cellobiose salicin	glucose	162
Trichoderma E-58	β-glucosidase	calcium alginate beads	cellobiose salicin	glucose	163
Porphyridium cruentum				sulphated polysaccharide	164

13

Table 2.1 (continued)

(I) Mechanical Attachment

(D) Entrapment in Polysaccharides

Agar

Cell	Microbial enzyme	Solid support	Substrate	End Result	Ref.
Saccharomyces pastorianus	invertase	agar pellets	sucrose	hydrolysis	38
Clostridium butyricum		agar gel	alcohol factory waste water	hydrogen	39
Rhodospirillum rubrum Klebsiella pneumoniae		entrapped agar	glucose	hydrogen	40
Clostridium butyricum		acetyl-cellulose filter with agar	peptone, glucose, riboflavin	hydrogen	41
Pseudomonas putida		agar gel	caffeine		43
Brevibacterium flavum		agar, acetyl-cellulose	glucose, urea, penicillin	glutamic acid	165
		agar gel, poly-acrylamide gel, collagen membrane, entrapped	wastewaters	methane	42
Azotobacter chroococcum	nitrogenase	agar gel		nitrogen fixation, γ-aminobutyrate, glutamate, arginine	166
Lactobacillus arabinosus		glass electrode, agar gel	nicotinic acid		167

(I) Mechanical Attachment

(D) Entrapment in Polysaccharides

Agar (cont'd)

Cell	Microbial enzyme	Solid support	Substrate	End Result	Ref.
Chloroplasts, Clostridium butyricum, spinach chloroplasts		agar gel		hydrogen	168
B. subtilis, Pseudomonas, L. bulgaricus, L. thermophilus		agar gel	milk	proteolysis	169
Anabena N-7363		agar gel		hydrogen	170
Chlorella vulgaris, Providencia sp.		agarose	amino acids	α-ketone acids	45
Bacillus subtilis, contg. plasmid pPCB6		agarose beads	L-broth, novobiocin	proinsulin	46
permeabilized Catharanthus roseus	hexokinase/ glucose-5-phosphate dehydrogenase	agarose			47
Catharanthus roseus		agarose, agar, κ-carrageenan, alginate, gelatin	sucrose	ajmalicine	48

15

Table 2.1 (continued)

(I)   Mechanical Attachment

(D)   Entrapment in Polysaccharides

Carrageenan

Cell	Microbial enzyme	Solid support	Substrate	End Result	Ref.
Escherichia coli	aspartase	carrageenan and locust bean gum, glutaraldehyde		L-aspartic acid	49
Escherichia coli		κ-carrageenan	ammonium fumarate	L-aspartate	50
Brevibacterium flavum	fumarase	κ-carrageenan and polyethyleneimine		L-malic acid	51
Brevibacterium flavum	fumarase	κ-carrageenan modified with amines			52
Yeast		κ-carrageenan, tricalcium phosphate			53
		agar, carrageenan, polyacrylamide			54
Brevibacterium flavum	fumarase	κ-carrageenan		ethanol	55
Saccharomyces cerevisiae		κ-carrageenan			56
Aspergillus, Candida, A. niger		polyacylamide, carrageenan, calcium arginate	beet molasses	citric acid	171

(I) __Mechanical Attachment__

(D) __Entrapment in Polysaccharides__

Carrageenan (cont'd)

Microbe	Microbial enzyme	Solid support	Substrate	Product	Ref.
Acetobacter aceti		carrageenan globules	apple juice, yeast extract	acetic acid	172
Lactobacillus thermophilus		κ-carrageenan	corn steep liquor, yeast extract, peptone, glucose	L-lactic acid	173
mutant Candida tropicalis		κ-carrageenan	n-alkane, n-alcohol, n-monoacid, α, ω-diol	α,ω-dodecanedionic acid, α, ω-tridecane-dionic acid	174
Enterobacter aerogenes		κ-carrageenan	glucose	2,3-butanediol	175
yeast		κ-carrageenan gel beads	glucose	ethanol	176
Corynebacterium glutamicum		κ-carrageenan hardening agent		glutamate	177
Pseudomonas dacunhae	L-aspartate β-decarboxylase	carrageenan glutaraldehyde	L-aspartic acid	L-alanine	178
Serratia marcescens		carrageenan		L-isoleucine	179
Brevibacterium fuscum		κ-carrageenan	dehydrocholic acid	12-ketochenodeoxy-cholic acid	180
Penicillium urticae, Conidia		κ-carrageenan beads			181

Table 2.1 (continued)

(I) Mechanical Attachment

(D) Entrapment in Polysaccharides

Carrageenan (cont'd)

Cell	Microbial enzyme	Solid support	Substrate	End Result	Ref.
Trichoderma reesei		κ-carrageenan		cellulase	182
Escherida coli	glutathione synthetase, acetate kinase	carrageenan, glutaraldehyde, hexamethylenediamine		glutathione	183
Actinoplanes missouriensis, Kluyveromyces fragilis, Saccharomyces cerevisiae	glucose isomerase, β-galactosidase, invertase	cellulose and cellulose di- and Tri-acetate beads			32
Actinoplanes missouriensis		cellulose beads	glucose	isomerization	34
Actinoplanes missouriensis		cellulose fibre	glucose	fructose	35
Sarcina ureae	urease	cellulose tri-acetate fibres	urea	hydrolysis	36
β-galactosidase, Kluyveromyces fragilis		cellulose and cellulose acetates entrapment	DMSO, acetone formamide, N-ethylpyridinium chloride		31
Actinoplanes missouriensis	glucose isomerase	cellulose beads, glutaraldehyde	glucose	isomerization	184

18

(I) <u>Mechanical Attachment</u>

(D) <u>Entrapment in Polysaccharides</u>

    <u>Alginate</u>

Cell	Microbial enzyme	Substrate	Solid support	End Result	Ref.
<u>Myobacterium Pyl</u>	NAD(P)H monooxygenase	alkene gas	alginate	epoxides	185
<u>Mastigocladus laminosus</u>		ADP + P_i + light	calcium alginate	ATP	186
<u>Digitalis lanata</u>			entrapped alginate	cardiac glycosides	187
spores <u>Curvularia lunata</u>		cortexolone	alginate	cortisol	188
<u>Arthrobacter simplex</u>	steroid-Δ^1-dehydrogenase	cortisol	alginate	prednisolone	189
<u>Pleurotus ostreatus</u>		penicillin	entrapped chitosan	6-aminopenicillanic acid	190
<u>Streptomyces</u> sp., <u>S. tendae</u>			calcium alginate gel	tylosin, nikkomycin	191
<u>Penicillium chrysogenum</u>			calcium alginate gel vesicles, entrapment	penicillin G	192
<u>Penicillium chrysogenum</u>, <u>Bacillus sp.</u>, <u>Bacillus subtilis</u>			calcium alginate gel polyacrylamide collagen	penicillin G., bacitracin, α-amylase	193
<u>Gluconobacter oxydans</u>		glucose	calcium alginate entrapped beads	gluconic acid	194

Table 2.1 (continued)

(I) Mechanical Attachment

(D) Entrapment in Polysaccharides

Alginate (cont'd)

Cell	Microbial enzyme	Solid support	Substrate	End Result	Ref.
Providencia sp.	L-amino acid oxidase	entrapped alginate	L-methionine	α-keto acids, α-keto-γ-methiol butyric acid	195
Trigonopsis variabilis	D-amino acid oxidase	entrapped alginate	amino acids	α-keto acids	196
Lactobacillus sp., Lactobacillus vaccinostercus		calcium alginate	xylose	lactic acid	197
Leuconostoc oenos		calcium alginate	L-malic acid	L-lactic acid	198
Lactobacillus delbrueckii		entrapped calcium alginate beads	glucose	L-lactic acid	199
Lactobacillus delbrueckii		entrapped calcium alginate beads	glucose, yeast extract, calcium carbonate	L-lactic acid	200
Saccharomyces cerevisiae		occlusion in pectin gel		ethanol	201
yeast		membrane alginic or pectic acid calcium alginate	molasses	ethanol	202
Saccharomyces cerevisiae		entrapped calcium alginate beads	polycation	ethanol	203

20

(I) <u>Mechanical Attachment</u>

(D) Entrapment in Polysaccharides

Alginate (cont'd)

Microbe	Microbial enzyme	Solid support	Substrate	Product	Ref.
Saccharomyces cerevisiae, Kluyveromyces marxianus		calcium alginate gels	glucose	ethanol	57
Kluyveromyces fragilis		alginate	whey lactose	ethanol	204
Pachysolen tannophilus		calcium alginate	D-xylose	ethanol	205
yeast		alginate	worts	beers	206
Saccharomyces cerevisiae		calcium alginate	glucose	ethanol	207
Saccharomyces cerevisiae		alginate $CaCl_2$		ethanol	208
Schizosaccharomyces pombe 077, Leuconostoc mesenteroides IAM 1233		entrapped alginate	malic acid	red wine	209
Zymomonas mobilis		alginate	glucose	ethanol	210
Pachysolen tannophilus		alginate gel	D-xylose	ethanol	211

21

Table 2.1 (continued)

(I)   Mechanical Attachment

(D)   Entrapment in Polysaccharides

Alginate (cont'd)

Microbe	Microbial enzyme	Solid support	Substrate	Product	Ref.
Saccharomyces cerevisiae		alginate	glucose, cane molasses	ethanol	212
Saccharomyces cerevisiae, Saccharomyces uvarum, Zymomonas mobilis		calcium alginate	glucose	ethanol	213
Saccharomyces fragilis		crosslinked alginate	lactose	ethanol	214
Kluyveromyces marxianus		alginate beads	Jerusalem artichoke tubers	ethanol	215
Saccharomyces cerevisiae		aluminum alginate	glucose	ethanol	216
Zymomonas mobilis		alginate κ-carrageenan		ethanol	217
β-glucosidase, yeast		entrapped alginate	cellobiose	ethanol	218
Zymomonas mobilis		calcium alginate	glucose	ethanol	219

22

(I) Mechanical Attachment

(D) Entrapment in Polysaccharides

Alginate (cont'd)

Cell	Microbial enzyme	Solid support	Substrate	End Result	Ref.
Clostridium acetobutylicum spores		calcium alginate gel		butanol	220
Clostridium acetobutylicum		alginate	glucose, acetone	BuOH	221
Saccharomyces cerevisiae		entrapped calcium alginate	glucose	ethanol	222
Methanosarcina barkeri		alginate cross-linked, entrapped	methanol	methane	223
Erwinia rhapontici		entrapped alginate	sucrose	isomaltulose	224
Saccharomyces cerevisiae		κ-carrageenan	sodium sulphite	glycerol	225
Pseudomonas denitrificans		alginate gel	ground water, ethanol, nitrate	denitrification	226
Candida tropicalis		aluminum alginate	phenol	degradation	58

23

Table 2.1 (continued)

(I) Mechanical Attachment

(D) Entrapment in Proteins

   Collagen

Cell	Microbial enzyme	Solid support	Substrate	End Result	Ref.
Streptomyces phaeochromogenes	glucose isomerase	collagen			65
Escherichia coli.		collagen membranes glutaraldehyde	fumaric acid	L-aspartic acid	69
Erwinia herbicola					70
Chlorella		collagen crosslinking dialdehyde starch soln.			71
Erwinia herbicola		collagen		tyrosine	227
Citrobacter freundii		collagen fibrils glutaraldehyde	cephalosporins		72
Clostridium butyricum		polyacrylamide gel membrane, agar gel, collagen	glucose	hydrogen	73
					228
Corynebacterium simplex		collagen, glutar-aldehyde membrane	hydrocortisone		135

(I) __Mechanical Attachment__

(D) __Entrapment in Proteins__

    __Gelatin__

Cell	Microbial enzyme	Solid support	Substrate	End Result	Ref.
Saccharomyces cerevisiae		cationic gelatin sawdust			75
Proteus mirabilis		formaldehyde gelatin	2-keto acids	2R-hydroxy acids	74
Streptomyces aureofaciens		gelatin crosslinked glutaraldehyde	daunomycinone	13-dihydrodauno- mycinone II	229
Saccharomyces cerevisiae	invertase	tuff granules gelatin	sucrose	hydrolysis	76
Saccharomyces cerevisiae	invertase	gelatin entrapped	sucrose	sucrose inversion	230
Saccharomyces cerevisiae	invertase	gelatin			231
Escherichia coli	lactase, invertase, catalase	hen egg white, glutaraldehyde crosslinking entrapped			77
Caldariella acidiphila	β-galactosidase	albumin glutaraldehyde entrapped			78

Table 2.1 (continued)

(I) Mechanical Attachment

(D) Entrapment in Polysaccharides

(c) Entrapment in Synthetic Polymers

  Polyacrylamide

Cell	Microbial enzyme	Solid support	Substrate	End Result	Ref.
S. cerevisiae + B. ammoniagenes		polyacrylamide gel	NAD + glucose + P_i	NADP	100
Achromobacter aceris	NAD kinase	polyacrylamide gel		NADP	96
Brevibacterium ammoniagenes	polyphosphate NAD kinase	polyacrylamide gel	org. solvents detergents	NADP	98
Saccharomyces cerevisiae	alcohol dehydrogenase	entrapped poly-acrylamide gel		NAD(H)	99
Brevibacterium ammoniagenes	NAD kinase	PGD, poly(vinyl alc.) and N,N'-methylenebis-acrylamide		NADP	97
Streptomyces fradiae		polyacrylamide gel	meat extract + starch	protease	232
Streptomyces fradiae		polyacrylamide gel		protease	85
Bacillus subtilis		polyacrylamide gel	meat extract + yeast extract	α-amylase	84

26

(I) Mechanical Attachment

(D) Entrapment in Polysaccharides

(c) Entrapment in Synthetic Polymers

Polyacrylamide (cont'd)

Cell	Microbial enzyme	Solid support	Substrate	End Result	Ref.
Curvularia lunata		polyacrylamide gel	Reichstein compound S	cortisol	233
Arthrobacter simplex	steroid-Δ¹-dehydrogenase	polyacrylamide gel	cortisol	prednisolone	86
Saccharomyces cerevisiae		acrylamide gel		steroid M-ketone reduction	234
Arthrobacter globiformis, Saccharomyces cerevisiae	3-ketosteroid-Δ¹-dehydrogenase	entrapped poly-acrylamide gel	steroids	steroids trans-formations	101
Pseudomonas testosteroni	Δ¹-dehydrogenase	polyacrylamide gel	Reichstein's substance S	steroid dehydrogenation	235
Arthrobacter simplex		polyacrylamide gel	hydrocortisone	prednisolone	236
Mycobacterium phlei			sterols 4-cholesten-3(o-car-boxy methyl)-oxime	side chain degradation	237

Table 2.1 (continued)

(I) Mechanical Attachment

(b) Entrapment in Polysaccharides

(c) Entrapment in Synthetic Polymers

Polyacrylamide (cont'd)

Cell	Microbial enzyme	Solid support	Substrate	End Result	Ref.
Arthrobacter globiformis	3-ketosteroid-Δ^1-dehydrogenase	entrapped polyacrylamide	hydrocortisone	transformations	83
Bacillus sp.		polyacrylamide gel	starch bouillon	bacitracin	238
Bacillus sp.		polyacrylamide gel	peptone	bacitracin	102
Streptomyces clavuligerus		crosslinked entrapped polyacrylamide		cephalosporin	103
Kluyvera citrophila		polyacrylamide gel	6-aminopenicillanic acid + D-phenylglycine methyl ester, peptone	ampicillin	88
Penicillium chrysogenum		polyacrylamide gel	glucose	penicillin G	239
Escherichia coli	β-lactamase	polyacrylamide	β-lactam antibiotics	degradation	240
Escherichia coli	aspartate ammonia lyase	polyacrylamide gel	ammonium fumarate	L-aspartic acid	241
Escherichia coli		polyacrylamide gel	ammonium fumarate	L-aspartate	242
Escherichia coli		reinforced polyacrylamide gel	ammonium fumarate	L-aspartic acid	93

28

(I) Mechanical Attachment

(D) Entrapment in Polysaccharides

(c) Entrapment in Synthetic Polymers

Polyacrylamide (cont'd)

Cell	Microbial enzyme	Solid support	Substrate	End Result	Ref.
Escherichia coli	aspartase	reinforced polyacrylamide gel	ammonium fumarate	L-aspartic acid	243
Escherichia coli		polyacrylamide gel	ammonium fumarate	L-aspartate	244
Escherichia coli		polyacrylamide gel	ammonium fumarate	L-aspartic acid	245
Escherichia coli	aspartase	polyacrylamide gel	ammonium fumarate	L-aspartic acid	246
Escherichia coli		polyacrylamide beads		L-tryptophan	247
Escherichia coli		entrapped poly-acrylamide gel	indole, pyruvate, ammonia	L-tryptophan	248
Escherichia coli	tryptophanase	polyacrylamide gel	indole, ammonia, pyruvate	L-tryptophan	249
Escherichia coli	tryptophan synthetase	polyacrylamide	indole, L-serine	L-tryptophan	250
Corynebacterium dismutans		polyacrylamide entrapment	glucose ammonia	alanine	251

Table **2.1** (continued)

(I) Mechanical Attachment

(D) Entrapment in Polysaccharides

(c) Entrapment in Synthetic Polymers

Polyacrylamide (cont'd)

Cell	Microbial enzyme	Solid support	Substrate	End Result	Ref.
Corynebacterium glutamicum		polyacrylamide		glutamic acid	87
Bacillus sp.		polyacrylamide gel	hydantoins	N-carbamyl-D-amino acids	252
Pseudomonas putida	L-arginine deiminase	polyacrylamide gel	L-arginine	L-citrulline	82
Escherichia coli	glutathione synthetase, acetate kinase	carrageenan gel		glutathione	183
Saccharomyces cerevisiae		polyacrylamide gel	L-glutamate + L-cysteine + glycine	glutathione	253
Saccharomyces cerevisiae		polyacrylamide gel	glucose, amino acids	glutathione	254
Candida lipolytica		polyacrylamide gel		citric acid	94
Candida lipolytica		polyacrylamide gel		citric acid	95
Brevibacterium ammoniagenes	fumarase	polyacrylamide gel	fumaric acid	L-malic acid	255
Saccharomyces cerevisiae	glucose oxidase	concanavalin A, bound to wall of Saccharomyces	sucrose	fructose, gluconic acid	256

30

(I) _Mechanical Attachment_

(D) _Entrapment in Polysaccharides_

(c) _Entrapment in Synthetic Polymers_

   Polyacrylamide (cont'd)

Cell	Microbial enzyme	Solid support	Substrate	End Result	Ref.
Saccharomyces cerevisiae, Aspergillus niger	invertase, catalase, glucose, oxidase	polyacrylamide γ irradiation	sucrose	gluconic acid fructose	257
Rat liver mitochondria	fumarase	polyacrylamide gel entrapment using gamma rays	fumaric acid	L-malic acid	258
Propionibacterium shermanii		polyacrylamide	glucose, $MgSO_4$ Na phosphate	carboxylic acid	259
Gluconobacter melanogenus		polyacrylamide gel	L-sorbose	L-sorbosone	260
Lactobacillus bulgaricus, E. coli, Kluyveromyces lactis		polyacrylamide gel	lactose	glucose + galactose	261
Achromobacter butyri		polyacrylamide gel	glucose metaphosphate	glucose-6-phosphate	262
Escherichia freundii		polyacrylamide gel	glucose + p-nitro-phenyl phosphate	glucose-6-phosphate + glucose-1-phosphate	263
Saccharomyces cerevisiae		entrapped crosslinked polyacrylamide	glucose	ethanol	264
Alcaligenes faecalis	β-glucosidase	entrapped polyacrylamide	cellulosic carbohydrate	ethanol	265

Table **2.1** (continued)

(I) Mechanical Attachment

(D) Entrapment in Polysaccharides

(c) Entrapment in Synthetic Polymers

Polyacrylamide (cont'd)

Cell	Microbial enzyme	Solid support	Substrate	End Result	Ref.
Saccharomyces cerevisiae		crosslinked prepolymerized polyacrylamide-hydrazide	glucose	ethanol	266
Candida tropicalis		polyacrylamide gel	phenol	degradation	123
Pseudomonas putida		polyacrylamide gel	benzene	degradation	80
Genus Arthrobacter		polyacrylamide gel	phenanthrene 3,4-benzpyrene	phenanthrene 3,4-benzpyrene	267
Acetobacter xylinum		polyacrylamide gel	glycerol	dihydroxyacetone	268
Clostridium butyricum		polyacrylamide gel membrane entrapped		hydrogen	269
Lactobacillus casei		polyacrylamide gel			270
Escherichia coli	L-aspartase	polyacrylamide gel			271
Rhizobium japonicum		entrapped polyacrylamide gel			272
Lactobacillus casei	malic acid decarboxylase	polyacrylamide gel			273

32

(I) Mechanical Attachment

(D) Entrapment in Polysaccharides

(c) Entrapment in Synthetic Polymers

Urethane

Cell	Microbial enzyme	Solid support	Substrate	End Result	Ref.
Arthrobacter simplex		urethane prepolymers	hydrocortisone 4-androstene-3,17-dione	prednisolone 4-androstene-3,17-dione	104
Corynebacterium sp.		entrapped photo-crosslinkable resin prepolymers, urethane prepolymers, polysaccharides	4-androstene-3,17-dione	9α-hydroxylation steroid	107
Nocardia rhodocrous		entrapped urethane	3β-hydroxy-Δ^5-steroids	3-keto-Δ^4-steroids	275
Nocardia rhodocrous		urethane prepolymers	testosterone	4-androstene-3,17-dione, dehydrotestosterone (DTS)	109
Nocardia rhodocrous		urethane prepolymers	3β-hydroxy-Δ^5-steroids	3-keto-Δ^4-steroids	110
Arthrobacter simplex		maleic poly-butadiene gel	hydrocortisone	prednisolone	111
Escherichia coli	aspartase	polyurethane	fumarate	L-aspartic acid	112
Caldariella acidophila	β-galactosidase	polyurethane foam entrapped	lactose	β-galactose	113

Table 2.1 (continued)

(I) <u>Mechanical Attachment</u>

(D) <u>Entrapment in Polysaccharides</u>

(c) <u>Entrapment in Synthetic Polymers</u>

<u>Urethane</u>

Cell	Microbial enzyme	Solid support	Substrate	End Result	Ref.
<u>Enterobacter</u> <u>aerogenes</u>		photo-crosslinkable resin prepolymer urethane prepolymer	uracil arabinoside adenine	adenine arabinoside	114
<u>Rhizopus delemar</u>	lipase	urethane prepolymers on silica beads	olive oil	inter-esterified fat	276
<u>Propionibacterium</u> <u>freudenreichii</u>		polyurethane			277
<u>Rhodotorula</u> <u>minuta var.</u>		photo-crosslinked	dl-menthyl succinate	l-menthol	278

34

(I) Mechanical Attachment

(D) Entrapment in Polysaccharides

(c) Entrapment in Synthetic Polymers

Photo Crosslinkable Resins

Cell	Microbial enzyme	Solid support	Substrate	End Result	Ref.
Hansenula jadinii		polyethylene glycol hydroxy- ethylacrylate		CDP-choline	279
Trichoderma reesei		radiation poly- merization 2-hydroxyethyl acrylate		cellulase	280
Citrobacter freundii	cephalosporinase	glass tubes photo-crosslinkable resin prepolymer	cephalosporin C	assay	281
Curvularia lunata		photo-crosslinkable resin			282
Nocardia rhodocrous	steroid-Δ^1- dehydrogenase	entrapped hydro- philic or lipo- philic gels	steroid	Δ^1-dehydrogenation	283
Arthrobacter simplex	Δ^1-dehydrogenase	entrapped photo- crosslinkable resin prepolymers	hydrocortisone	prednisolone	117

35

Table 2.1 (continued)

(I) Mechanical Attachment

(D) Entrapment in Polysaccharides

(c) Entrapment in Synthetic Polymers

Photo Crosslinkable Resins

Cell	Microbial enzyme	Solid support	Substrate	End Result	Ref.
Rhizopus stolonifer spores		polysaccharides entrapped photo-crosslinked resin prepolymers, urethane prepolymers	progesterone potato-dextrose broth	11α-hydroxylation	284
Arthrobacter simplex Mitochondria		photo-crosslinkable resins, entrapped	hydrocortisone	prednisolone	285
Rhodotorula minuta, Nocardia rhodochrous		poly(propylene glycol), urethane	cholesterol and dehydroepiandrosterone	cholestenone and 4-androstene-3,17-dione	286
Curvularia lunata		entrapped photo-cross-linked resin gels, mycelium, polyethylene glycol	cortexolone	hydrocortisone	287
Curvularia lunata		photo-crosslinkable resin prepolymer	Reichstein's compound S	hydrocortisone	288

36

(I) Mechanical Attachment

(D) Entrapment in Polysaccharides

(c) Entrapment in Synthetic Polymers

Photo Crosslinkable Resins

Cell	Microbial enzyme	Solid support	Substrate	End Result	Ref.
Corynebacterium sp.		photo-cross-linkable resin	4-androstene-3,17-dione	9α-hydroxylation	107
Escherichia coli	L-threonine deaminase	photo-cross-linkable resin prepolymer	L-threonine	analysis	289
Clostridium sp.		entrapped photo-crosslinkage resin	Δ^2enoates or aldehydes with hydrogen gas	hydrogenation	290
Kloeckera sp.	alcohol oxidase, catalase, D-amino acid oxidase	photo-cross-linkable resin	methanol, D-amino acid	oxidase	116

Table 2.1 (continued)

(I) Mechanical Attachment

(D) Entrapment in Polysaccharides

(c) Entrapment in Synthetic Polymers

    Miscellaneous

Cell	Microbial enzyme	Solid support	Substrate	End Result	Ref.
Enzymes or bacterial cell bodies		polystyrene, glycidylmethacrylate			291
Escherichia coli		polystyrene beads	rose bengal		120
Candida fumicola	D-amino acid oxidase	methacrylate			119
Streptomyces phaeochromogenes		radiation-induced polymer of 2-hydroxyethyl methacrylate trapped			236
Zymomonas mobilis		entrapped borosilicate glass fibre pads, flocculation		ethanol	148
Yeast		polyvinyl alcohol on bentonite	glucose, $MgSO_4 \cdot 7H_2O$	ethanol	124
Saccharomyces cerevisiae		entrapped silica hydrogel	glucose		126
Saccharomyces cerevisiae		amine-cured epoxy resin		ethanol	123

38

(I) Mechanical Attachment

(D) Entrapment in Polysaccharides

(c) Entrapment in Synthetic Polymers

Miscellaneous

Cell	Microbial enzyme	Solid support	Substrate	End Result	Ref.
Arthrobacter simplex	corticosteroid dehydrogenase	surfactants polybutadiene			125
Escherichia coli		epoxy		6-amino-penicillanic acid	122
Streptomyces phaeochromogenes		2-hydroxyethyl methacrylate polymer	glucose	fructose	118

Table 2.1 (continued)

(II) Covalent Attachment

Cell	Microbial enzyme	Solid support	Substrate	End Result	Ref.
Saccharomyces cerevisiae		alginate, poly-ethyleneimine + glutaraldehyde or carbodiimide + N-hydroxysuccinimide or periodate, poly-ethyleneimine			63
Escherichia coli yeast	lactase invertase catalase	hen egg white, glutaraldehyde			77
Zygosaccharomyces lactis		covalently linked hydroxyalkyl meth-acrylate, epichlor-hydrin, amine spacer			292
Saccharomyces paradoxus		hydroxyalkyl meth-acrylate gel, covalent glutaral-dehyde			137
organelles and whole cells		protein foam, serum albumin and glutar-aldehyde			136
Caldariella acidophila	β-galactosidase	albumin and glutaraldehyde			78
Desulfovibrio desulfuricans	periplasmic hydrogenase	glutaraldehyde, calcium alginate, polyacrylamide(s)			293

40

(2) Covalent Attachment

Cell	Microbial enzyme	Solid support	Substrate	End Result	Ref.
Saccharomyces paradoxus		hexamethylene diamine, hydroxy-alkyl methacrylate gel, glutaraldehyde			294
Saccharomyces cerevisiae		covalent hydroxyalkyl methacrylate gel			295
Yeast		polyphenylene oxide glutaraldehyde			296
Acetobacter suboxydans		polyacrylamide gels glutaraldehyde	alditols, aldose diethyl dithioacetals		297
Chlorella		collagen, dialde-hyde, dialdehyde starch soln.			109
Catharanthus roseus	isocitrate dehydrogenase, cathenamine reductase	agarose, poly-urethane, glutar-dialdehyde, hexa-methylenediamine			298
Bacillus coagulans	glucose isomerase	glutaraldehyde			299
Citrobacter freundii		collagen fibrils glutaraldehyde	cephalosporins		72
Peroxisomes from Kloeckera		albumin crosslinked with glutaraldehyde	hydrocortisone	prednisolone, catalase, D-amino acid oxidase, and alc. oxidase	285

41

Table 2.1 (continued)

(2) Covalent Attachment

Cell	Microbial enzyme	Solid support	Substrate	End Result	Ref.
Streptomyces clavuligerus		polyacrylamide, acylhydrazide groups, glutardialdehyde, polyvinyl alcohol		cephalosporins	103
Solanum aviculare		adsorption covalent linkage activated polymeric adsorbent glutaraldehyde		steroid glyco-alkaloids	300
Corynebacterium simplex		collagen glutaraldehyde	hydrocortisone	prednisolone	135
Escherichia coli		collagen membranes glutaraldehyde	hydrocortisone	prednisolone	69
Escherichia coli		carrageenan and locust bean gum glutaraldehyde and hexamethylene-diamine		L-aspartic acid	49
Escherichia coli	glutathione synthetase and acetate kinase	carrageenan glutaraldehyde hexamethylenediamine		glutathione	183
Pseudomonas dacunhae		glutaraldehyde	L-aspartic acid	L-alanine	178
Protaminobacter rubrum		glutaraldehyde flocculation	sucrose	isomaltulose	301

(2) Covalent Attachment

Cell	Microbial enzyme	Solid support	Substrate	End Result	Ref.
Proteus mirabilis		formaldehyde, gelatin	2-keto acids	2R-hydroxy acids	74
Clostridium sp.		formaldehyde gelatin or polyacrylamide gel	2-enoates	saturated acids	
Saccharomyces uvarum		open pore gelatin calcium alginate glutaraldehyde	cane molasses	ethanol	302
Saccharomyces cerevisiae		glutaraldehyde tannin		wine	133
Yeast/cellulase		tannin glutardialdehyde	cellobiose	ethanol	132
Micrococcus luteus	ammonia-lyase	carboxymethyl-cellulose carbodiimide	L-histidine	urocanic acid	303
Bacillus subtilis α-amylase		zirconia-coated alkylamine glass with glutaraldehyde			304
Saccharomyces uvarum Schizosaccharomyces pombe Bacillus megaterium		glass, brick beads glutaraldehyde			305

Table 2.1 (continued)

(III) Ionic Attachment

(A) Flocculation

Cell	Microbial enzyme	Solid support	Substrate	End Result	Ref.
Yeast, Schizo-saccharomyces pombe		flocculation	malt extract yeast extract bactopeptone	ethanol	145
Sporasarcina urea		flocculation Na bentonite			139
Aggregate		flocculated polyelectrolyte	glucose	fructose	306
Protaminobacter rubrum		cation-active flocculant glutaraldehyde	sucrose, syrup, corn steep liquor	isomaltulose	301
Zymomonas mobilis		glass-fibre pads		ethanol	148
Yeast		Ca alginate			147
Glucose Isomerase		flocculation by cationic and anionic polyelectrolytes	glucose	glucose isomerase fructose	7
Stemphylium loti	cyanide hydratase	mycelia	cyanide	formamide	307
Yeast		Ni, Fe_3O_4 and $CaCO_3$ particles			142
Saccharomyces cerevisiae		Ni powder, Fe sand flocculation			141
Streptomyces No. 41	glucose isomerase	coagulated chitosan			143
Nitrosomonas europaea		trimethylammonium glycol chitosan iodide and potassium			144

(III) Ionic Attachment

(B) Adsorption

Cell	Microbial enzyme	Solid support	Substrate	End Result	Ref.
Acetobacter sp.		titanium (IV) hydroxide	ethanol	acetic acid	308
Acetobacter sp.		hydrous titanium(IV) oxide	ethanol	acetic acid	152
Escherichia coli Acetobacter		hydrous TiO$_2$	wort	malt vinegar	309
Aspergillus niger		mycelia, glass		gluconic acid citric acid	310
Pseudomonas paucimobilis		soil	glucose	biomass	311
Saccharomyces carlsbergensis		porous glass γ-aminopropyltri-methoxysilane (I) glutaraldehyde	glucose	CO$_2$, EtOH	312
Yeast		ceramic			313
Arthrobacter globiformis	3-ketosteroid Δ^1-dehydrogenase	silica gel CrCl$_3$ or TiCl$_3$	hydrocortisone	prednisolone	314
Anabaena cylindrica		glass beads		hydrogen, oxygen	315
Streptococcus pyogenes, Neisseria gonorrhaea, Haemophilus influenzae		adsorbed cordierite, alumina			316

45

Table 2.1 (continued)

(III) Ionic Attachment

(B)   Adsorption

Cell	Microbial enzyme	Solid support	Substrate	End Result	Ref.
Saccharomyces cerevisiae		inert carrier	cellulose hydrolysate glucose	ethanol	317
yeast, Saccharomyces cerevisiae		colloidal Al(OH)$_3$ spheres, colloidal metal oxide or hydroxide glass plate			318
Clostridium thermocellum		bituminous coal particles	cellulose, biomass	alcohol	319
Acetobacter aceti		cordierite		acetic acid	320

46

(B) **Adsorption**

On polysaccharides and ion exchangers

Cell	Microbial enzyme	Solid support	Substrate	End Result	Ref.
Saccharomyces cerevisiae		beechwood chips	sugar	ethanol	321
Yeast		sugar cane bagasse pitch	sugar	ethanol	322
Yeast				ethanol	323
Saccharomyces cerevisiae		cellulose, ECTEOLA-cellulose films	sugar/tuber hydrolysate	ethanol	324
Pseudomonas aeruginosa		corn stover	nitrates and nitrites		325
Pseudomonas aeruginosa		cellulose fibres	nitrate	nitrite	326
Azotobacter vinelandii	nitrogenase	Cellex E	N_2	N fixation NH_3	327
Azotobacter vinelandii		Cellex E			328
Nocardia erythropolis		adsorbed on DEAE-cellulose	cholesterol	cholest-4-ene-3-one	153
Nocardia erythropolis, N. opaca, Mycobacterium phlei	steroid 1-dehydrogenase	adsorbed to DEAE-cellulose	3-oxo-23,24-bisnor-1,4-choladienic acid	3-oxo-23,24-bisnor-1-cholenic acid	329
Saccharomyces cerevisiae		ion exchange resins		ethanol	154

The dried cell aggregate retained all its enzyme activity in glucose isomerization [5, 6]. Other applications of mycelial aggregates in fermentation include a kinetic study using *Morterella vinacea* mycelial pellets in p-nitrophenyl-α-D-galacto-pyranoside hydrolysis from the action of α-galactosidase [7] and the use of myce-lium of *Absidia griseola* in the production of the enzyme α-galactosidase free of invertase [8]. Mycelium pellets of *Aspergillus ochraceus* were used to optically re-solve N-acetyl-DL-methione to ℓ-methionite [9].

By means of special techniques, the filamentous growth of a fungus can be en-couraged to produce mycelial masses. Thus by innoculating the mycelium of a fungus and using the polymer Carbopol-934 to control growth, fungal mycelial pellets of uniform size can be obtained [10]. Addition of the complexing agent 18-crown-6(1,4,7,10,13,16-hexaoxacyclooctadecane) to the culture medium was found to produce cells of up to 200 times normal length for rod shape species such as *Escherichia coli, Salmonella typhimurium*, and *Bacillus sp.* [11].

Use of these filamentous cells together with a membrane with high porosity is proposed as a means of facilitating fermentation [12]. Special particles with a cage structure formed by folding or crushing small sheets of wire meshes have been used to retain bacterial filamentous fungal organisms for fermentation reactions [13]. The most important application of mycelial aggregates in fermentation, the production of acetic acid, has been reviewed by CONNER and ALLGEIER [14].

Mycelial filaments together with their constituent enzymes can also be isolated by flocculation and bound to solid supports by various methods of immobiliza-tion such as adsorption, encapsulation, and entrapment. These methods are dis-cussed separately since immobilization in these instances does not involve pre-dominantly mechanical forces.

Fermentation using enzymes contained in mycelial mats and pellets as well as in microbial cell walls is one of the oldest methods and is applicable mostly in the preparation of bulk products and in waste treatment where enzyme/cell contam-ination is not critical. Accumulation of mycelial mass may cause problems in a fermentation reaction.

2.1.2 Encapsulation

Encapsulation of enzymes has been reviewed by CHIBATA [15] who describes four existing procedures: interfacial polymerization, liquid drying, phase separation and liquid membrane methods. Two of the methods, the liquid drying and liquid membrane methods, have been used successfully in microbial systems.

In these methods, microbial cells are completely surrounded by a layer of semi-permeable material such as collodion, silicone, collagen, cellulose acetate, poly-ester, nylon, or ethylcellulose. The molecular dimensions of the microcapsules im-pose a limit on the molecular sizes of both nutrients and products in fermentation. There is also a dependence of the reaction rate on the rate of transfer of substrate into the capsules. For example, using tracers with a range of molecular weights, a nylon microcapsule membrane was estimated to transmit molecular weights in the range 1000 to 10000 [16]. The dimensions of the capsule will also limit the growth of the cells, although an initial high cell density is possible.

An example of liquid drying is encapsulation of the yeast Saccharomyces cerevisiae in ethylcellulose [17]. Ethylcellulose was dispersed in a benzene/n-hexane mixture by the dispersing agent sorbitan monolaurate. To the chilled mixture was added a suspension of *S. cerevisiae* in 1% NaCl solution to form an emulsion, which was then further emulsified with a polyethylene glycol/water solution. To this emulsion was added n-hexane to precipitate ethylcellulose microcapsules of diameters 100–300 μm. The microcapsules retained up to 71% of the enzyme activity for 5–10 days and, compared with the entrapment method using polyacrylamide gel, immobilized 4–5 times more microorganisms per unit gel volume and resulted in a 4–5 times higher enzyme activity. However, the liquid dried immobilized cells were significantly weaker than entrapped cells. Some improvement in this microencapsulation procedure was obtained by employing chitosan pretreatment.

Another variation of the technique of solid encapsulation is the liquid membrane method in which the cells, for example, viable strains of *Micrococcus denitrificans* ATCC 21909, are added to a surfactant mixture to give an emulsion containing droplets of 20–40 μm in diameter each enclosing 500–600 cells in the presence of a membrane strengthener and an anion transport facilitator [18]. It was found that, after 120 h, 78% of the initial enzyme activity was retained, and that after 5 days of operation, there was no cell lysis. The reaction rate was found to be dependent on the rate of substrate transfer, a fact that can be utilized in maintaining a degree of selectivity in the diffusion of substrates into the capsules by the choice of different liquid membranes. Other advantages observed are the high capacity of catalyst regeneration and the stability of the membranes. One disadvantage in liquid membrane encapsulation was pointed out by CHANG [19], namely the tendency of sedimentation to occur inside the capsules.

The advantages of the encapsulation method are the mechanical and chemical stability of the membrane system, the possibility of high loading, and regulation of the fermentation reaction by selective diffusion of substrates and products [20]. By the same token, some substances may have too low a rate of diffusion.

2.1.3 Dialysis Culture

Dialysis culture includes a wide range of examples, most of which belong to the field of immobilization in principle only, although there is considerable overlap in their classification. Thus in dialysis culture in fermentation, some cells are confined or localized physically by a diffusion barrier through which nutrients and fermentation products up to a selected molecular size can freely pass. Microbial cells may also be bound to such a barrier and examples of these kinds of immobilization are covered later.

Dialysis culture has been reviewed in depth by SCHULTZ and GERHARDT [21] and more recently by ABBOTT [22]. SCHULTZ and GERHARDT identify four modes of operation in dialysis fermentation. These modes are represented by a fermentor containing the cells interfaced through a dialysis membrane with a reservoir containing the nutrient and serving as a sink for the dialyzed fermentation product; either or both of the two vessels can be operated batchwise or continuously. The pore size of the membrane ranges from 25 to 200 nm (filter membranes, permeable to solutes and macromolecules), 0.3 to 10 nm (dialysis membranes, permeable to sugars and salts) to solution transport membranes (permeable only to gases). Mathematical models for the observed dialysis and growth phenomena in these four modes of operation were proposed for reactor design. SCHULTZ and GERHARDT pointed out that up to 1969 there had been little application of dialysis fermentation in the production of metabolites, although they observed and cited examples [23, 24] to support the fact that in fermentation where product inhibition exists, the use of a dialysis membrane is beneficial because it keeps the products away from the microorganisms. A recent example of this technique is the high biomass concentration (140–150 g/l) of E. coli obtained by LANDALL and HOLME [25, 26] from which the growth inhibiting products, the low molecular weight carboxylic acids, were excluded by dialysis.

Dialysis fermentation found a slightly wider application by 1978, when ABBOTT [22] provided five examples claiming yield improvement over conventional batch methods by factors varying from 2 to 45. Despite the advantages of in dialysis fermentation, the method has not been used in industrial production. The main reasons for this are:

1. Low economic potential for larger scale industrial production.
2. Some fermentation products cannot be dialyzed.
3. Possibility of leakage of some microorganisms or cell fragments by dialysis.
4. The problem of membrane fouling.

2.1.4 Entrapment

By far the most frequently used technique of cell immobilization, entrapment methods call for incorporation of the cells within the lattice of a polymeric material, which can be carbohydrate, protein, synthetic organic or inorganic. The common enzyme immobilizing technique in which the solid carrier is in the form of a membrane is also used to immobilize cells by entrapment.

Carbohydrates
Two kinds of polysaccharides are most frequently used in entrapping cells: cellulose and its derivatives, and the gel-like substances extracted from sea-weeds – for example, agar from *Gelidium corneum* and related algae, alginate from the brown sea-weed *Macrocysts pyrifera* (L.) Ag., *Lessoniaceae,* carrageenan from the red algae *Chondrus cripsus* and *Gigantina stellata,* and finally pectin from skins of citrus fruits. The alga polysaccharides are more stable towards alkaline than towards acid conditions and are normally insoluble in alcohol or alcohol/water mixtures. They are reasonably stable towards heat, up to 100 °C in water, at which they can be extracted from the alga. Dilute aqueous solutions (up to a few percent) when chilled or when their counter ions are replaced by calcium, as in the case of alginates, become gelatinous solids in which water, together with the cells, is trapped inside a network of polysaccharide molecules. As a rule these gels have little mechanical strength, especially towards abrasion, but can be strengthened by crosslinking with bifunctional reagents such as glutaldehyde.

Polysaccharides such as agar, alginate and carrageenan have the important feature that they resemble the physical environment found in a microbial cell and as such confer stability to entrapped cells and enzymes. However, their mechanical properties are not outstanding; agar is unstable towards high temperature and calcium alginate is unstable in the presence of chelating agents such as phosphate salts. Cellulose and its derivatives are not soluble in water, but dissolve in polar aprotic organic solvents such as dimethyl formamide, dimethyl sulphoxide and acetone. Whole microbial cells can also be added to these solutions. When the solution is passed into water it can be drawn as fibres or formed into beads and membranes containing the immobilized microorganisms.

Cellulose and Derivatives. Immobilization of microbial cells in a cellulose matrix follows procedures established earlier for enzymes [27, 28]. An early example of entrapping whole cells in cellulose triacetate fibres is provided by DINELLI [29] based on procedures developed for enzymes. To a standard wet spinning apparatus was added an emulsion formed by mixing a cellulose triacetate solution in methylene chloride and a chilled cell suspension in a water/glycerol buffer. This emulsion was then extruded through a spinneret by nitrogen pressure into a chilled bath of toluene to coagulate. The fibres were then collected and vacuum dried. DINELLI found that *Escherichia coli* cells immobilized on cellulose triacetate fibres in this manner at a concentration of 1 mg wet cells/g polymer have 80% of the enzyme activity (penicillin acylase) of the free cells. However, cells of *Saccharomyces lactis* entrapped in a similar way but at a higher concentration of 75 mg dry cells/g polymer showed only 10% of the galactosidase activity of the free cells.

Among the early attempts to entrap whole cells in a matrix of cellulose derivatives may be cited the work of KOLARIK et al. [30]. A solution of cellulose triacetate in methylene chloride was mixed with an aqueous suspension of whole cells containing the enzyme glucose isomerase, stirred vigorously and a secondary solvent added. The product was made into fibres (250×500 µm) by injection through a syringe into toluene, or by casting into membranes (10–20 µm thick) on glass or water. Membranes with entrapped microorganisms were found to retain 57% of the original cell activity, although there was a problem with cell leakage. New methods of dissolving cellulose or its derivatives were devised, using solvents such as dimethylsulphoxide, acetone-formamide and a melt formed with N-ethylpyridinium chloride and dimethylsulphoxide [31].

LINKO [32] used a modified procedure to entrap whole cells in cellulose beads. In this procedure, a solution of cellulose or its derivatives (α-cellulose in a N-ethylpyridinium chloride/dimethylsulphoxide melt; cellulose diacetate in dimethyl sulphoxide/acetone or cellulose triacetate in dimethyl sulphoxide) was prepared at 90 °C and then cooled to 30 °C, to which was added dried microbial cells to form a cell suspension. After deaeration, this suspension was extruded into water at 23 °C to form cellulose beads which can be hardened by crosslinking with glutaraldehyde. Physical parameters affecting performance include bead size, porosity, cell load, mechanical strength. Microorganisms immobilized by this method [33] include *Saccharomyces cerevisiae* (invertase) with a half life of 5 years, *Kluyveromyces fragilis* (β-galactosidase), and *Actinoplanes missouriensis*, on which kinetic data were obtained from a plug-flow column reactor utilizing glucose isomerase activity. Data on reactor kinetics and mass transfer were obtained from immobilized cells in a packed bed reactor [34]. *Actinoplanes missourienses* whole cells containing the active enzyme glucose isomerase, entrapped in α-cellulose fibres and cross-linked with glutaldehyde, have been used in the preparation of fructose from glucose [35]. The extent of conversion was about 40% and the process in a packed glass column was operated continuously for 23 days. Studies were carried out on whole *Sarcina ureae* cells entrapped in cellulose triacetate fibres [36], based on the urea hydrolysis by the cell enzyme urease at pH 7.2 and 60 °C with a half-life of 177 h at a residence time of 1.32 h. It was observed that both the free and the entrapped cells had the same optimum temperature and pH.

Other examples of entrapment of whole cells in cellulose are shown in Table 2.1. In general, this technique provides high loading and operational stability [37]. The flow rate can be optimized by the proper choice of polymer carrier, bead size and the physical characteristics of the carrier [32].

Agar. Long used as a culture medium for microorganisms, agar is a logical choice for a solid carrier in cell immobilization. It is readily soluble in boiling water and solidifies at 37 °C. TODA and SHODA [38] entrapped whole cells of *Saccharomyces pastorianus* containing the enzyme invertase in spherical pellets of agar and studied the kinetics of sucrose hydrolysis in a continuously fed fluidized bed reactor. In a typical procedure, agar solution 2.5% (w/v) and a yeast cell suspension were mixed in the ratio of 4:1 by volume at 50 °C. The mixture was then injected through a 36 mm tube into a chilled solution of toluene and tetrachlorethylene to form agar pellets, which were then sieved through acrylic resin screens to obtain particles of a uniform size. Microscopic examination confirmed that the distribution of yeast cells inside the spherical pellets was uniform.

SUZUKI et al. [39] found that immobilization of the hydrogen producing microorganism *Clostridium butyricum* in 2% agar gel had a stabilizing effect on the system of enzymes of the whole cell, especially hydrogenase, so that continuous production of hydrogen is possible. To a solution of agar in physiological saline cooled to 50 °C was added another solution, in physiological saline, of wet whole cells of *C. butyricum*. After cooling to 37 °C, the solid gel was cut up, washed, and used in continuous production of hydrogen from alcohol production on waste water over a 20 day period at a rate of 6 ml/kg wet gel. A similar improvement in stability was reported by WEETALL et al. [40] in an agar-entrapped mixture of *Rhodospirillum rubrum* and *Klebsiella pneumoniae* which showed an operational half-life of about 1000 h, producing hydrogen by photometabolic means in a reactor using the substrate glucose obtained from hydrolysis of paper and sawdust.

KARUBE et al. [41] immobilized growing cells of *Clostridium butyricum* in an acetylcellulose carrier with agar and found that addition of flavic compounds such as peptone, and, in particular, riboflavin, enhanced both the number of cells and the production of hydrogen from the substrate glucose, this effect being especially noticeable after the third repeated use. Routine fermentation using resting cells instead of growing cells produced only 10–30% of the stoichiometric amount of hydrogen based on glucose. The exact mechanism of the flavin effect is not known. KARUBE et al. [41] cited the participatory role of the flavins in hydrogen and electron transfer in the enzyme (oxidoreductase or dehydrogenase) catalyzed reaction.

Because of its non-toxicity, agar gel may be a superior carrier for methanogenic bacteria compared to other carriers such as collagen/glutaraldehyde and those based on acrylamide polymerization, which exert an inactivating, toxic effect on the cells [42]. Other applications of agar entrapped microbes include decaffeination [43] by *Pseudomonas putida* and preparation of amino acids by the protoplast of *Brevibacterium flavum* [44]. The bio-decaffeination of coffee, using the immobilized enzyme extracts of *Pseudomonas putida,* in contrast with the existing chemical extractions using methylene chloride or carbon dioxide, calls for regeneration of the coenzyme NADH. To avoid this difficult step, whole growing caffeine-resistant cells of *P.putida* were entrapped in agar gel. Up to 99% removal of the caffeine was reported. However, several problems still remain, among which is the non-selective degradation of other constituents of coffee and a requirement for arsenate as ATP-ase inhibitor. In the second example, cells of *Brevibacterium flavum* were first lysed to release the protoplasts, added to a buffered solution of $MgCl_2$, agar (3%) and acetyl cellulose, then chilled to solidify. Glutamic acid was produced from a substrate of glucose, urea and penicillin. Other applications of immobilization in agar are listed in Table 2.1.

Agarose. Agarose is a neutral gelling fraction obtained from agar. It is soluble in hot water and solidifies when a 1% solution is cooled. A procedure for immobilizing microbial cells in agarose was developed [45] in which the cell suspension was first mixed with a 5% agarose solution at 45 °C, then dispersed in vegetable oil at the same temperature. The droplets (diameter 1 mm) were solidified by chilling to 5 °C, then filtered and washed to remove oil. By this procedure, cells of the alga *Chlorella vulgaris* and the bacterium *Providencia sp.* were co-immobilized and used in the oxidation of amino acids to α-keto acids. The algal cells, when illuminated with red light, produced the oxygen required in the conversion and accounted for a tenfold rate increase.

A second application of the agarose procedure was found in the production of proinsulin. After several solid carriers, including alginate, polyacrylamide and agarose, were tested for their characteristics in immobilizing cells of *Bacillus subtilis* carrying plasmids encoding rat proinsulin and in releasing the proteins (proinsulin) synthesized in the ensuing fermentation, MOSBACH et al. [46] selected 2% (w/v) agarose to which 5 $\mu g/ml^{-1}$ of novobiocin, a DNA gyrase inhibitor, was added to suppress cell division in favour of protein production.

Cells of the plant *Catharanthus roseus,* simultaneously permeabilized with various organic solvents for rapid diffusion of the secondary metabolites produced in a fermentation, were entrapped in agarose, producing ajmalicine isomers from cathenamine [47, 48].

κ-Carrageenan. The use of κ-carrageenan for whole cell immobilization was initiated by NISHIDA et al. [49] for the continuous production of L-aspartic acid by *Escherichia coli.* A general procedure for the immobilization is to mix at 40 °C a suspension of the cells in 0.9% sodium chloride solution with a solution of κ-carrageenan in 0.9% sodium chloride. The mixture is cooled to 10 °C for 30 min, then treated with chilled 0.3 M KCl solution. The resultant gel is finely divided, and further treated with cross-linking reagents such as glutaraldehyde or a mixture of glutaraldehyde and hexamethylenediamine, giving entrapped cells with aspartase activity of 37 460–49 400 units/g cells.

Using the immobilized whole cell preparation of *E. coli* in packed column reactors, SATO et al. [50] studied the reaction parameters in terms of continuous production of L-aspartic acid and found that the immobilized cells at the optimum pH level of 9.0 and at 55–60 °C (vs. pH 9.5 and 50 °C for the free enzyme) gave MICHAELIS constants 5 times higher than the free enzymes. L-aspartic acid was produced at 90% yield.

In the production of L-malic acid by κ-carrageenan entrapped *Brevibacterium flavum* it was found [51] that the presence of polyethyleneimine (0.15%) had such desirable effects as increased stability and activity of the cell enzyme and increased productivity compared to the untreated immobilized cells. An attempt was made to determine the mechanism of this stabilization effect [52] by reproducing the stabilization effects using κ-carrageenan containing amino substituent groups. The existence of a "tripartite" interaction between κ-carrageenan, polyethyleneimine and *B. flavum* cells was suggested, although the exact mechanism is still not clear. Other stabilization reagents include tricalcium phosphate (hydroxyapatite) [53], polyacrylamide [54], glutaraldehyde or a mixture of glutaraldehyde and hexamethylene diamine [49].

Among the beneficial effects of these additions, the most obvious ones are the improved mechanical characteristics of the solid carrier – increased hardness, porosity, leak proofing, stability towards heat, abrasion, shear forces, extreme pH levels – and enhanced enzyme stability and activity. Using κ-carrageenan immobilized whole cells of *Brevibacterium flavum*, TAKATA et al. [55] demonstrated that the order of stability towards such operational parameters as temperature, agitation, solvent action, and the denaturing action of chemical reagents was first the immobilized whole cells (the most stable), then the free cells and finally the native enzyme (the least stable). The stabilizing effect was more pronounced when the carrier was in the gel state than in the solid state. The cell leakage encountered in fermentation with κ-carrageenan entrapped cells was investigated by WANG et al. [56] who demonstrated that it can be traced to mechanical causes such as agitation rate and shear forces. This conclusion was supported by the observation that the rate of leakage is constant for carriers containing homogeneously entrapped cells. The rate of leakage can be lowered by hardening the carrier in a solution containing potassium cations. Examples of applications using κ-carrageenan entrapped whole cells are given in Table 2.1.

Immobilization on κ-carrageenan has been used to prepare a number of chemicals ranging from alcohols, organic acids, amino acids to enzymes and antibiotics (Table 2.1). In many cases, immobilization used in conjunction with the hardening agent glutaraldehyde confers high carrier stability and long microbial half-lives. Because of the favourable conditions available inside the κ-carrageenan gel lattice (so that rigorous incubation conditions are not required), this carrier is widely used to entrap growing or viable microorganisms. Generally fermentations using this carrier produce less by-product and a higher yield of the desired products.

Alginate. Alginic acid, a mixture of D-monouronic acid and L-guluronic acid, is obtained from the giant sea weed *Macrocystis pyrifera* (L.) *Ag. Lessoniaceae* and other sea weeds. The acid form, alginic acid, is soluble in alkaline solutions but is insoluble at pH <3. Sodium alginate absorbs 200–300 times its weight of water and is insoluble in alcohol, or aqueous alcohol containing more than 30% (w/w) of water. When the monovalent counter ion of sodium is replaced by divalent calcium, ionic crosslinking among the carboxylic acid groups occurs and the polysaccharide molecules form a polymeric network in which about 60% water is entrapped to give a gelatinous substance, calcium alginate. Because the structural characteristics depend on the action of the calcium ion, reagents that form stable complex compounds with calcium (most phosphates and chelating agents such as ethylenediaminetetraacetic acid) will mechanically weaken the gel. Microbial cells entrapped in calcium alginate usually show enhanced operational stability and activity.

As an example, to entrap a sample of microbial whole cells, for example *Saccharomyces cerevisiae* [57], the wet cells were harvested and mixed with a 1% solution of sodium alginate in water. This mixture was taken up in an automatic pipette with a tip opening of 1 mm and slowly exuded into a 0.05 M $CaCl_2$ solution containing 10% by weight of glucose. The insoluble fibres were collected and packed into a column reactor. The nutrient medium consisted of a continuous flow of a 0.05 M $CaCl_2$ solution containing 10% glucose, from which the ethanol content is monitored. The coagulating solution of $CaCl_2$ was replaced by an aluminum chloride solution containing the cells in suspension, into which was dropped a solution of sodium alginate [58]. In comparing the mechanical properties of alginate gels made with calcium, barium, and strontium divalent cations, in the immobilization of *Rhodopseudomonas capsulata* chromatophores, showing photophosphorylating activity, PAUL and VIGNAIS [59] found that the bacterial chromatophores entrapped in barium alginate retained up to 70% of their activity and showed better physical-chemical properties than in calcium alginate. Physical properties of calcium alginate gels have been studied by CHEETHAM [60] and CHEETHAM et al. [61].

Other attempts to stabilize calcium alginate gels towards the action of phosphates involve formation of cross-linking polymers. Cationic polymers such as polyethyleneimine and polypropyleneimine when used to cure calcium alginate beads in aqueous solution apparently [62] replace calcium ions and produce a shrinking effect on the gel. Treatments of polyethyleneimine longer than 10 min caused a decrease in respiration rate in the entrapped *Escherichia coli* cells, although the beads showed stability towards a 0.1 M K_2HPO_4/KH_2PO_4 solution at pH 7.5. In addition to the ionic bonding in calcium alginate, crosslinking can also be effected by covalent bonding [63], using polyethyleneimine and glutaraldehyde. Gel-entrapped *Saccharomyces cerevisiae* cells treated by covalent crosslinking were reported to have the same activity and viability as untreated cells, and to be stable in buffered 0.1 M sodium phosphate solution with respect to leakage of cells. However, in many instances, especially when crosslinking is applied after cell immobilization on alginate, intercellular and cell to lattice covalent bonding will occur. This method of immobilization is discussed later under covalent bonding.

The viability of plant cells entrapped in calcium alginate gel was demonstrated [64] for *Daucus carota, Cannabis sativa*, and *Ipomoea sp.*, which were found to remain viable for 8–24 days. Alginate has been used extensively as a carrier in fermentations involving a large number of immobilized microbial and plant cells. By far the predominant application is in the production of ethanol as a solvent and in wine and beer making. Butanol and glycerol are also produced by this method. Fermentations are usually conducted continuously in packed column or fluidized bed reactors, both of which have an advantage over stirred tank reactors in that damage to the carrier from mechanical forces is less likely. Of the organic acids, lactic acid is produced most frequently by using alginate as the carrier for several species of *Lactobacillus* cells. α-Keto acids are made from the corresponding α-amino acid by bacterial oxidation. There are several examples involving preparations of antibiotics, steroids, cardiac glycosides and ATP (Table 2.1).

Miscellaneous. Other polysaccharides which have been used in entrapment of whole cells include pectin and chitosan (Table 2.1). Pectin occurs in citrus fruit rinds and is a partial methyl ester of copolymers of D-galacturonate and L-rhamnose. It is soluble in water and gives a solution when chilled. As an ester it is not stable towards strong acid or base solutions. Chitosan is a cationic polysaccharide derived from chitin which forms the exoskeletons of marine invertebrates, anthropods and fungi. Chitin, an acetamido derivative of a poly$(1\rightarrow4)$2-dioxy-β-D-glucose, is deacetylated by hydrolysis with base to give chitosan.

Protein

Collagen. Collagen, a fibrous protein, occurs in animal connective tissues and has been used as a support for tissue culture and also as casing for sausages. Collagen fibrils form an open polar, hydrophilic structure which provides numerous sites for entrapping cells and enzymes. Owing to its fibrous nature, collagen can be cast into membranes which can readily be adapted to a number of reactor configurations for fermentation.

VIETH et al. [65] describe a procedure for immobilizing in collagen whole cells of *Streptomyces phaechromogenes* containing the enzyme glucose isomerase. A suspension of the frozen cells in water is first heated to 80 °C for 1.5 h, cooled to room temperature and added to a collagen solution initially at pH 6.5. Additions of NaOH solution raises the pH to 11.2 to disperse the collagen solution which is then cast into membranes of 2–10 mils thick. The membranes can be strengthened by crosslinking with glutaraldehyde. The purpose of the heating at 80 °C is to fix the enzymes thermally. The technology of collagen membrane immobilized whole cells and enzymes has been reviewed by VENKATASUBRAMANIAN [66], VIETH and VENKATASUBRAMANIAN [67], and KOLOT [68]. Based on experimental evidence, the nature of the bonding action, at least in the case of enzymes, is considered to be a combination of ionic forces such as hydrogen bonding, van der Waal's attractions and salt linkages [67]. However, such studies are not available for immobilized cells and it may be conjectured that in this case due to the much larger size of the cells, bonding by purely mechanical confinement inside the fibrous structure of collagen may play a more important role. The technique has been applied to the production of citric acid by viable cells using all the enzymes connected with the TCA cycle [67]. Half-lives were of the order of 150–300 h and relative productivity 14–48% of the traditional fermentation using free cells.

Crosslinking a collagen membrane with glutaraldehyde confers to the membrane superior mechanical properties and helps prevent cell leakage. VENKATASU-BRAMANIAN et al. [69] describe the advantages of such a technique in fermentations using several microorganisms to produce L-aspartic acid and glutamic acid. However, the use of glutaraldehyde probably has a detrimental effect on some microorganisms so that the viability of the immobilized cells cannot be ascertained [69]. In one instance, methanogenic bacteria from waste water were shown to be inactivated by glutaraldehyde and oxygen in a collagen membrane but remained growing and active in an agar gel matrix [42]. The use of the crosslinking agent dialdehyde starch was demonstrated to be superior to glutaldehyde for *Erwinia herbicola* cells in the yield of L-tyrosine [70]. This crosslinking agent forms the basis of a patent for microbial cell immobilization on collagen membranes [71]. Collagen membranes with entrapped microbial cells are also used in biological sensors for substances such as cephalosporin [72] and oxygen [73].

Gelatin and Albumin. When the collagen in animal connective tissues is treated with acid or base and then extracted with boiling water, gelatin is obtained. In the solid state, gelatin absorbs water and on heating disperses in the aqueous medium as a colloidal solution, which can reversibly solidify when chilled. Gelatin is often used in connection with the crosslinking agents glutaldehyde or formaldehyde [74] and often with an inert support such as saw-dust [75] or tuff [76].

Entrapment of yeast and *Escherichia coli* cells on egg albumin was attempted by D'SOUZA et al. [77] in conjunction with permeabilization by toluene. This carrier was found to be superior for whole cells of *Caldariella acidophila* [78].

Synthetic Polymers

Polyacrylamide and Polyurethane. Polyacrylamide is the synthetic polymer most often used in immobilization of microbial cells. Polyacrylamide, formed by the linear polymerization of the vinyl carboxylic acid amide, is a solid readily soluble in water. The polymer is normally crosslinked to the copolymer N,N'-methylenebisacrylamide to produce a polymer with lattice-like structure better suited to cell immobilization.

An early example of immobilization of viable microbial cells in polyacrylamide is the work of UPDIKE et al. [79], who reported that viable cells of the protozoan *Tetrahymena pyriformis* and the bacterium *Escherichia coli* could be immobilized in a polyacrylamide gel. The immobilization procedure involves treating a solution of the monomers acrylamide and N,N'-methylenebisacrylamide in phosphate buffer (pH 7.4) with the solution of the catalysts, N,N,N',N'-tetramethylenediamine, sodium hydrosulphite, and riboflavin. Polymerization was initiated at 20 °C by a photo flood-lamp and the resulting gel broken up and sieved to −40 mesh. Viability of the cells up to 5 days was demonstrated by oxygen uptake, glucose utilization and mobility. Insight into the activity and viability of polyacrylamide-entrapped *Pseudomonas putida* cells was obtained through electron microscopy [80]. It was observed that there was a 40–70% drop in the activity of the immobilized cells, and that the cells could be reactivated by incubating them in a medium with benzene and succinate or with iron in the absence of a carbon source. Electron microscope observation demonstrated that the initial entrapped cells, existing in ill-defined colonies on the carrier, soon grew on incubation in a suitable medium into spherical colonies containing an increased number of cells. Continuous elution caused these spherical colonies to disintegrate, exposing cells with ruptured cell walls. On reactivation, the cell population was restored.

The main advantages of using polyacrylamide as carrier to entrap whole microbial cells are the resulting stabilization effects on the cell-related enzymes. The cell-related enzymes are found to be more stable towards γ-irradiation [81] or heat [82].

Because they are more exposed, intracellular enzymes allow higher rates of diffusion from the cell [82]. Activities of these enzymes after immobilization are usually equal to [83, 84] or greater than [81, 85–88] those of the free cells. This stabilizing effect on the enzyme due to immobilization is also evident from the durability of enzyme activity on repeated cycles of use [84, 89]. The viability of cells entrapped in polyacrylamide is closely tied to their thermal stability since the polymerization is an exothermic reaction. For example, entrapped variants of *Arthrobacter globiformis* lost 93.6% of their viability on standing for 20 min at 4–10 °C or 99.5% on standing for 20 min at 24 °C [83]. The thermal stability of polyacrylamide entrapped *Alcaligenes faecalis* was rated between that of the free cells and that of the free enzymes and the cells retained 85% of their activity after being immobilized in polyacrylamide and a cross-linking agent at 30 °C [90, 91]. In this respect, radiation induced polymerization, using 2-hydroxyethyl methacrylate, for example, is a better alternative, especially in view of the excellent adhesion reported with *Streptomyces phaeochromogenes* [92]. Another adverse effect of polyacrylamide as a carrier comes from the toxicity of the reagents of the polymerization reaction, namely, the initiator, the crosslinking reagents and the solvent system. For example, the promotor/initiator mixture of $S_2O_3^=$ and $Me_2N(CH_2)_2CN$ was found to lower the activity of *A. faecalis* by 20% in 30 min [90]. The mechanical strength of un-reinforced polyacrylamide is not entirely suitable for fermentation reactors in which abrasion and shear is expected, for example, in a stirred or in a fluidized bed reactor [93]. Polyacrylamide has been found [94, 95] to be unsuitable for the production of citric acid by immobilized cells of *Candida lipolytica*. There is constant leakage of cells, low citrate production and the presence of undesirable side products.

Polyacrylamide is used in the preparation of fine chemicals such as amino acids, steroids, antibiotics, enzymes, coenzymes, and other nucleotides. Highly pure NADP has been produced by coimmobilizing NAD-kinase [96–98] and NADH by alcohol dehydrogenase [99]. NADP has also been prepared in a fermentation employing two coimmobilized cells, *Saccharomyces cerevisiae* and *Brevibacterium ammoniagenes* [100].

Fermentations involving steroids are mostly functional group transformations. KOSHCHEYENKO et al. [101] have studied a number of such transformations by the cells of *Arthrobacter globiformis*. Microbial fermentations on polyacrylamide gel produced high yields of the antibiotic bacitracin. Cells of *Bacillus chemother* [102] in a starch-bouillon medium showed evidences of growth as indicated by increase of activity and by observation with the electron microscope. Crosslinking reaction with glutaraldehyde, when carried out under physiological conditions, results in yields of cephalosporin similar to cells of *Streptomyces clavuligerus* [103]. The two amino acids most frequently obtained from fermentation of immobilized whole cells on polyacrylamide are L-aspartic acid and L-tryptophan.

Entrapment of whole cell and organelles in polyurethane was exploited by TANAKA et al. [104] who used prepolymers containing polyethylenediol units (average molecular weight 2600) with different ethylene oxide content (91% and 100%). The terminal groups of both prepolymers come from toluene diisocyanate. A chilled suspension of *Arthrobacter simplex* cells in a potassium phosphate buffer at pH 7.0 was mixed with a solution of the prepolymer. The mixture was poured onto a glass plate and kept at 4 °C for 1 h to gel. The polymer, after washing with phosphate buffer at pH 7, showed 15% of the free cell activity in transforming hydrocortisone to prednisolone. Similar procedures on *Nocardia rhodocrous* cells produced entrapped cells with 18% of the free cell activity in a similar transformation. FUKUI et al. [105] and FUKUI and TANAKA [106] have reviewed this topic.

Polyurethane is the carrier of choice in steroid fermentations, which are as a rule very sensitive to solvent and carrier polarity because of the hydrophobicity of steroid compounds. By judicious choice of the polarity and chain length [107] of prepolymers, especially the glycol components, it is possible to design a carrier for a particular fermentation. For example, cells of *Nocardia rhodocrous* entrapped in hydrophilic gels cannot oxidize the 3β OH group in cholesterol when nonpolar solvents are used [108]. The same cells entrapped in lipophilic gels produce 4-androstene-3,17-dione from testoterone by the enzyme Δ'-dehydrogenase while those entrapped on hydrophilic gels give dehydrotestosterone [109].

Enzymatic activities of the entrapped *N. rhodocrous* cells are higher in 3β-hydroxy oxidation [110], appreciable in steroid Δ'-dehydrogenation [104] and moderate in 1,4-dione formation from 4-androstene-3,17-dione [111]. Polyurethane carriers are used in fermentations of fumarate to L-aspartic acid [112], lactose to β-galactose [113], and uracil arabinoside and adenine to adenine arabinoside [114] with superior yields or enhanced enzyme activities.

Photocrosslinkable Prepolymers. These versatile polymer carriers have been developed by FUKUI et al. [105]. In general, they consist of two or three components. The simpler one, poly(ethyleneglycol)methacrylate (PEGM), polyethylene-glycol reacted by condensation with a methacrylate moiety at each end. The more complex polymer consists of three components, a central bifunctional polyethylene glycol of variable chain length attached at each end to another bifunctional molecule through its twin diisocyanate groups. The remaining diisocyanate group reacts with a hydroxyalkylacrylate.

Typical procedures for preparation of photocrosslinkable prepolymers may be illustrated by the procedure for poly(ethylene glycol)dimethacrylate [115]. The procedure involved refluxing a mixture of polyethyleneglycol and excess methacrylate in toluene with a trace of p-toluenesulphonic acid, heating at 70 °C a mixture of hydroxyethylacrylate and isophorone in the presence of an organotin catalyst, and then treating the reaction mixture with half the molar ratio of poly(ethylene glycol) at 70 °C for 5 h. This oligomer product was not further purified and is mixed with benzoin ethyl ether in water, melted at 50 °C and mixed with a yeast suspension in sucrose solution and a potassium sulphate buffer at pH 7.2. A layer of this mixture was illuminated at 300–400 nm for 3 min to form an insoluble gel which can be granulated for fermentation use. Another related prepolymer was prepared from polyethylene glycol, isophorone diisocyanate and hydroxyethylacrylate [116].

Photocrosslinkage prepolymers have the advantage of flexibility in design because components of different polarity and chain length can be built into the prepolymer. This flexibility is reflected in their frequent use in both the hydrophilic and the lyophilic forms in steroid fermentations, as in the case of the polyurethanes, in conjunction with solvents of different polarity, with a resulting improved enzyme stability. For example, when up to 30% of the main chain component polyethylene glycol was replaced by the more hydrophobic polypropyleneglycol, in the form of the prepolymer on which *Arthrobacter simplex* cells were immobilized, the hydrocortisone to prednisolone conversion increased [117].

Other Synthetic Polymers. Other acrylate derivatives used in photopolymerization include 2-hydroxyethyl methacrylate [118, 119] and 2-ethylhexyl acrylate [119] in a mixture with methyl methacrylate and acrylic acid.

Escherichia coli cells have been bonded to polystyrene beads [120]. Enzyme or bacteria have been entrapped in capsules formed from polystyrene and glycidylmethacrylate when treated at −78° with ethanol and hardened by irradiation with ^{60}Co. Treatment with acetone may soften the capsules by swelling up to 310% [121]. KLEIN and WAGNER [122] have demonstrated that epoxy beads have better characteristics (loading capacity, mechanical properties and versatility in rate control) than other solid carriers (alginate, polymethacrylate, polyurethane epoxide) for the entrapment of *E. coli* cells.

Amine-cured epoxy resin has also been shown to be suitable for ethanol production by viable cells of *Saccharomyces cerevisiae* with the possibility of continuous reincubation [123]. Polyvinyl alcohol [124], by itself or in a mixture with laminated clay bentonite, has been used to entrap *S. cerevisiae* cells to produce 94% of the theoretical yield of ethanol after 11 h in a column reactor containing a glucose substrate solution.

Arthrobacter simplex has been entrapped in polybutadiene by the action of UV light in the presence of an initiator and a photosensitizer [125].

Finally, inorganic polymers such as silica hydrogel [126], alkoxysilane [127] and borosilicate glass fibres [128] have been used in cell entrapment. These mineral carriers resist the mechanical degradation forces present in some fermenters and require only mild immobilization procedures. As an example, *Zymomonas mobilis* cells have been bonded to borosilicate in the fermentation of glucose to ethanol [128].

2.2 Covalent Attachment

The distinguishing feature of this method of immobilization is the existence of covalent bonds between the cells and the polymer lattice, or covalent bonds among the cells themselves to form a mat. Because of their defined chemical nature, enzymes are amenable to covalent bonding through bifunctional reagents. Use of such reagents on whole cells will usually kill the cells, although enzymatic activity may be preserved.

Immobilization by covalent bonding to enzymes has been reviewed extensively by CHIBATA [129]. Many reagents have been developed to bond the N–H, OH, SH groups as well as the imidazole group found in the polypeptide structure of enzymes.

Glutaraldehyde is the most commonly used crosslinking reagent to immobilize microbial cells. The dialdehyde reacts readily with amines to form a Schiff base

$$-NH = CH - CH_2CH_2CH_2 - CH = N-$$

in the condensation reaction and is often used to harden protein gels after immobilization.

Flocculated cells of *Protaminobacter rubrum* have been crosslinked with glutaraldehyde. When the attached cells were placed in a column reactor at 50° under a stream of 70% sucrose, no sucrose was detected in the outflow [130]. It has been proposed that glutaraldehyde be used to crosslink cells of *Bacillus coagulans* [131]. The cell wall of the yeast *Saccharomyces cerevisiae* has been used as a solid carrier to adsorb the enzyme cellulase [132] and pepsin [133] and then crosslinked by treating the cells with tannin and glutaraldehyde. A procedure has been proposed to mechanically strengthen the carrier calcium alginate containing entrapped *Saccharomyces cerevisiae;* the bonded carrier is stirred in a solution of polyethylene imine and calcium chloride for 24 h, washed, then stirred in a glutaraldehyde and sodium phosphate solution briefly so that the polyethylene imine, molecules of which diffuse and become ionically bonded to the calcium alginate lattice, becomes crosslinked covalently by reaction with glutaraldehyde [63]. A variation of this technique is to dissolve selectively in the glutaraldehyde treated carrier the portion that has a low degree of covalent crosslinking so that its mechanical strength as well as its porosity can be improved [134].

From the high stability of crosslinked carriers, BIRNBAUM [63] concluded that cells are covalently bonded by glutaraldehyde. Glutaraldehyde intercellular crosslinking is employed on many occasions when whole cells are entrapped in polysaccharide or synthetic organic polymer carriers (Table 2.1).

Covalent crosslinking with glutaraldehyde is particularly suitable for protein carriers such as collagen, gelatin, albumin, etc., most of which can be cast into a membrane or foam structure after loading and then crosslinked.

Covalent crosslinking between the green alga *Chlorella* and collagen is used in a patent [71] in which a cell suspension is mixed with a collagen solution and then extruded into an aqueous salt solution or an organic solvent followed by treatment with glutaraldehyde. Collagen membranes [69] containing whole cells of *Escherichia coli* crosslinked with glutaraldehyde showed a 95–98% fumaric acid to L-aspartic acid conversion. Other applications that have been demonstrated to be feasible are the production of monosodium glutamate [69], prednisolone [135] and the construction of a microbial sensor [72].

Two examples of glutaraldehyde crosslinking using *Proteus mirabilis* and *Clostridium* strain on gelatin have been given by TISCHER et al. [74]. Comparison with the respective free enzymes shows that immobilized cells are more stable under operational conditions.

The glutaraldehyde crosslinking method has been applied to the carrier albumin in a morphological observation by means of the transmission electron microscope [136] and in an investigative study on the microorganisms *Escherichia coli* (lactose activity) and yeast (calatase and invertase activities) giving rise to carrier systems with different permeabilities [77].

Covalent crosslinking is also possible on inorganic solid carriers such as silica, titania, glass, brick, and zirconia, usually with glutaraldehyde. Other means of covalent bonding, discussed by BIRNBAUM et al. [63], include pretreating alginate with

1. Polyethyleneimine then crosslinking with glutaraldehyde.
2. A carbodiimide and N-hydroxsuccinimide; the ester so formed then crosslinked with polyethyleneimine.
3. Periodate to cleave the glycol functions on alginate to dialdehydes, then crosslinking with polyethyleneimine.

Covalently bonded cells in general do not present problems of cell leakage or mechanical degradation. Feasibility studies concern the permeability, the operational stability of the enzymes and their activity, and the degree of loading and molecular make-up of the carrier [137]. Problems arise mainly from the loss of viability of the immobilized cells. In fact, the cell walls in most cases act as a solid carrier on which the enzymes are bound. Fermentations resorting to this method of immobilization are usually one-enzyme reactions since the same problems will be magnified greatly for multiple-enzyme reactions. Nevertheless, an attempt has been made to coimmobilize yeast with cellulytic enzymes to convert cellulosic material to ethanol in a single step [132].

2.3 Ionic Attachment

2.3.1 Flocculation

Mechanisms proposed to describe the ionic attachment interaction among microorganisms in the course of flocculation have been reviewed by FLETCHER [138]. The ionic attachment mode of immobilization makes use of predominantly electrostatic forces such as hydrogen bonding, coordinate bonding, Van der Waal and dispersion forces. When dispersed in water, hydrophobic particles acquire a surface change and hydrophilic particles a solvation layer, both acquired properties acting to repulse the particles and form a dispersion. Whether the dispersion

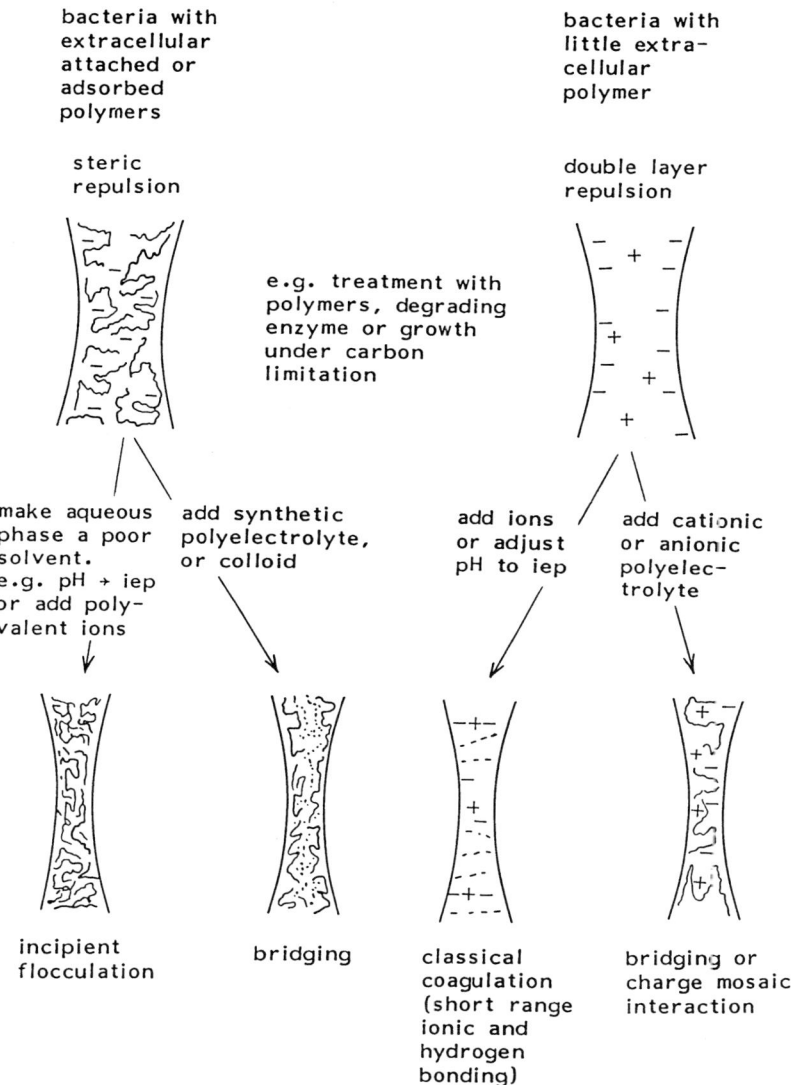

Fig. 2.1. Incipient flocculation and polymer bridging (i.e.p. = isoelectric point) [330]

is stable or coagulates is determined by the balancing of these forces of repulsion and short range forces of attraction.

By manipulating liquid phase conditions such as pH, ionic strength, or by addition of a flocculent to alter the counter-ion layer, it is possible to bring about conditions favouring short range attractive interactions between the particles so as to allow flocculation (Fig. 2.1). Microbial cell flocculates can then be used in the immobilized form in a fermentation. There are many mechanisms by which flocculation of microbial cells, which usually have negatively charged surfaces, may occur, among which are the use of multivalent ions Al^{3+} or Mg^{2+} either through a reduction in the counter ion layer around the cells because of the higher charge density of the multivalent ions, or through interaction with the counter ion layer to form hydroxides with metal ions as bridges. Although these bonds have a partial covalent character, the technique is classified as ionic attachment for convenience.

The mechanism of microbial flocculation is dependent on many factors as outlined by ALEMZADEH et al. [139]. These factors include the morphology and the physiology of the species as well as the conditions of the environment such as pH, ionic and nutritional characteristics, etc. Many reagents will induce flocculation in a microbial suspension: polyvalent cations such as Mg^{2+}, polyamine [140] or carboxyl-substituted polyacrylamide [140], minerals such as bentonite [139] and metallic particles such as iron and nickel. Metallic powders have been utilized [141] to flocculate yeast cells rapidly and effectively (up to 99.9%) with a rate independent of temperature and pH in contrast to an earlier study [142] of the same flocculation which was shown to contain inhibiting compounds. When the amine groups of chitosan are acidified [143] or methylated [144] the remaining ammonium groups in the polymeric structure are suitable for adsorbing cells on addition of an anionic polyelectrolyte [144]. The mechanism of adsorption is thought to be based on formation of salt bridges between the polyelectrolytes and the cell surface [144]. Flocculation is apparently caused by cell growth of *Schizosaccharomyces pombe* in a medium containing glucose, malt extract, yeast extract and bactopeptone [145]. More than 95% of the cell population is reported to be flocculated.

An extreme form of immobilization consists of heating the microorganism to fix the enzyme inside the cell so that there is no leakage and so that other activity is retained. The dead cells with the enzyme are then employed in a fermentation. The method has been described by CHIBATA et al. [146].

Flocculation is most often used in conjunction with other techniques in preparative fermentations. Microbial flocs that have low porosity can be mixed with diatomaceous earth to facilitate filtration [140]. The mechanical properties of flocs can be improved by crosslinking with glutaraldehyde or addition of alginate [147], but due to the susceptibility to degradation by abrasion, etc., flocs are usually used in fixed bed reactors [6].

Fermentations using flocs are usually one enzyme reactions; however, continuous generation of ethanol is reported [148] by flocs of organisms on glass fibre pads tentatively identified as *Zymomonas mobilis*. Another example is the fermentation of *Schizosaccharomyces pombe* flocs in a column reactor to produce ethanol [145]. In both cases, there is a connection between cell growth and ethanol production.

2.3.2 Adsorption

The phenomenon of microbial cell adhesion by adsorption is almost universal. GERSON and ZAJIC [149] define the thermodynamic work of adhesion in terms of thin surface free energies of the components of a bond. Materials such as mica and glass show strong but high surface energies and most fluorocarbons low surface energies.

On Minerals

Soil as a mineral is the most important carrier of microorganisms. The complex mechanisms of immobilization of microorganisms in soil have been reviewed by BURNS [150]. In fermentation work, various minerals have been used, namely, oxides of silicon (as in glass, gels, ceramics, clays), oxides of aluminium (clays, gels) and phosphates of calcium (apatite). Oxides of titanium, zirconium, aluminium, vanadium and iron can assume a hydrated structure in which the metallic elements, linked by hydroxy bridges, give rise to a polymeric compound. When some of these hydroxy bridges are replaced by similar substituents from the cell wall of a microorganism, immobilization occurs. Immobilization by adsorption on hydrous oxides has been reviewed by KENNEDY [151]. A typical procedure of immobilization calls for a solution of the titanium hydroxide as a mixture of its tetrachloride in dilute hydrochloric acid. The mixture is then neutralized with dilute ammonium hydroxide solution to pH 7.0. After repeated washing with dilute saline solution to remove the ammonium ions, the hydroxide is mixed with a suspension of *Escherichia coli* cells at pH 5–7 (pH 2 for other organisms), stirred at room temperature, allowed to settle, and the immobilized cells separated by centrifuging. Cells of *E. coli* immobilized in this way have been found to be unaffected by leaching with solutions of carbonate, phosphate, etc. and have shown 30% of the rate of oxygen uptake of the free cells. This method of immobilization has the advantages of easy preparation, and flexibility of pH range, from pH 7, which is physiologically suitable for most microorganisms, to as low as pH 2 [152].

Hydrous titanium oxides as carriers have the advantages of easy preparation and good attachment under physiological conditions. Living cells can be immobilized by this technique. Interaction between the oxide and the cells enables them to resist wash out at high flow rates and high loadings.

Other inorganic oxides may be used as adsorption carriers, among them silica (glass), ceramics, alumina, and the mineral cordierite, usually with retention of cell activity and viability.

On Polysaccharides

Cellulose in its various natural forms and its many derivatives is suitable for immobilizing microorganisms because of its non-toxicity, desirable mechanical properties, ready availability, and flexibility of structural and functional-group modifications. Derivatives of cellulose act as ion exchangers. Acidic substituents such as carboxymethyl (CM) and phosphoric acid (P) produce cationic exchangers, whereas basic groups such as diethylaminethyl (DEAE) and epichlorhydrin-methanolamine (ECTEOLA or Cellex E) produce anionic exchangers. Because of the overall negative changes in most microorganisms, anion exchangers are preferred in immobilization by adsorption.

In the oxidation of the steroid cholesterol to cholest-4-ene-3-one by immobilized cells of *Nocardia erythropolis*, DEAE cellulose has been shown to be a better carrier than polyacrylamide gel, alumina or agar [153]. In this application, moist cells of *N. erythropolis* are suspended in a phosphate buffer at pH 8.0, and stirred with a DEAE-cellulose suspension for 1 h at room temperature. The cells are separated, washed and used in fermentation in an organic solvent. Natural cellulose such as beech wood chips, sugar cane bagasse pith and corn stover have been frequently used as carriers for cell adsorption. In most cases, an increase is reported in the activity of the immobilized cells, or a superior yield of fermentation products and improved loadings. Because of the mild conditions, cells are usually viable metabolically and reproductively and have good operational stability, within certain pH and temperature ranges. The three most common areas of application are: steroid transformations, nitrogen fixation, and ethanol production, where the possibility of cell leakage or washout does not pose a great problem in purification.

On Ion Exchange Resins

DAUGULIS et al. [154] have compared the extents of adsorption of *Saccharomyces cerevisiae* cells on a number of anion exchange resins, among which the resin XE-352 (registered trademark, Rohm and Haas Canada Ltd) is very effective (128 mg dry cell/g dry resin). Non-polar adsorbents which have been used in immobilization include hydrocarbon and polystyrene.

Advantages of the adsorption method are the simplicity of operation, the mild conditions employed, the possibility of high cell concentration, long term stability and the minimization of cell build-up and sloughing [155]. However, adsorption is usually non-specific and highly dependent on pH and the ionic nature of the solvent. Early applications of ion exchange methods in immobilization have been reviewed by ROTMAN [156], who has concluded from adsorption of *Escherichia coli* on two different ion exchangers, both in the chloride form with drastically different degrees of adsorption, that, in addition to the presence of electrostatic charges in the bacteria, the configuration of the charges may play an important role in the adsorption.

References

1. Maddox IS, Dunnill P, Lilly MD (1981) Biotechnol Bioeng 23:345
2. Briffaud J, Engasser M (1979) Biotechnol Bioeng 21:2093
3. Briffaud J, Engasser JM (1979) Cell Immobilisees Colloq 197
4. Wiseman A (ed) (1983) Principles of biotechnology. Surrey University Press, New York, p 19
5. Nobuzo T, Takafumi K (1977) US Patent No 4,001,082, 4 January
6. Bungard SJ, Reagen R, Rodgers PJ, Wyncoll KR (1979) ACS Symposium Series 106, p 139
7. Kobayashi H, Suzuki H (1976) Biotechnol Bioeng 18:37
8. Narita S, Naganishi H, Yokouchi A, Kagaya I (1976) US Patent No 3,957,578, 18 May
9. Hirano K, Karube I, Suzuki S (1977) Biotechnol Bioeng 19:311
10. Van Suijdam JC, Kossen NWF, Paul PG (1980) Eur J Appl Microbiol Biotechnol 10:211
11. Tso W-W, Fung W-P (1981) Biotechnol Lett 3:421
12. Tso W-W (1980) Biotechnol Lett 2:519
13. Atkinson B, Black GM, Pinches A (1980) Proc Biochem 15:24
14. Conner HA, Allgeier RJ (1976) Adv Appl Microbiol 20:81
15. Chibata I (1978) Immobilized enzymes. Research and development. Wiley, New York, p 54

16. Miyawaki O, Nakamura K, Tano T (1980) Agric Biol Chem 44:2865
17. Sameyima H, Kimura K, Ado Y, Suzuki Y, Tadokoro T (1978) Enzyme Eng 4:237
18. Mohan RR, Li NN (1975) Biotechnol Bioeng 17:1137
19. Chang TMS (1971) Nature 229:117
20. Mohan RR, Li NN (1975) Biotechnol Bioeng 17:2
21. Schultz JS, Gerhardt P (1969) Bacteriol Rev 33:1
22. Abbot BJ (1978) Ann Rep Ferment Progr 2:91
23. Gerhardt P, Gallup DM (1963) J Bacteriol 86:919
24. Huang HT (1961) Appl Microbiol 9:419
25. Landall P, Holme T (1977) J Gen Microbiol 103:345
26. Landall P, Holme T (1977) J Gen Microbiol 103:353
27. Linko Y, Pohjola L, Linko P (1978) Bioconvers Food Technol, Proc Fr-Finn Symp Biotechnol 2nd, p 78
28. Linko Y, Poutanen K, Weckström L, Linko P (1979) Enzyme Microbiol Technol 1:26
29. Dinelli D (1972) Proc Biochem 7:9
30. Kolarik MJ, Chan BJ, Emery AH, Lim HC (1974) In: Olsen AC, Cooney CL (eds) Immobilized enzymes in food and microbial processes. Plenum Press, New York, p 71
31. Weckstrom L, Linko YY, Linko P (1980) Food Proc Eng 2:148
32. Linko P (1979) Food Proc Eng 2:27
33. Linko P, Poutanen K, Weckstrom L, Linko Y (1980) Biochemie 62:387
34. Linko P, Poutanen K, Linko Y (1981) J Mol Catal 13:263
35. Linko YY, Pohjola L, Linko P (1977) Proc Biochem 12:14, 32
36. Ghose TK, Kannan V (1979) Enzyme Microb Technol 1:47
37. Marconi W, Morisi F (1979) Appl Biochem Bioeng 2:219
38. Toda K, Shoda M (1975) Biotechnol Bioeng 17:481
39. Suzuki S, Karube I, Matsunaga T, Kuriyama S, Suzuki N, Shiogami T, Takamura T (1980) Biochemie 62:353
40. Weetall HH, Sharma BP, Detar CC (1981) Biotechnol Bioeng 23:605
41. Karube I, Urano N, Matsunaga T, Suzuki S (1982) Eur J Appl Microbiol Biotechnol 16:5
42. Karube I, Kuriyama S, Matsunaga T, Suzuki S (1980) Biotechnol Bioeng 22:847
43. Middlehoven WJ, Bakker CM (1982) Eur J Appl Microbiol Biotechnol 15:214
44. Azimoto Co. Inc. Japan Kokai Tokkyo Koho (1982) Japanese Patent No 57,144.989, 7 September
45. Wikström P, Szwajcer E, Brodelius P, Nilsson K, Mosbach K (1982) Biotechnol Lett 4:153
46. Mosbach K, Birnbeaum S, Hardy K, Davies J, Bülow L (1983) Nature 302:543
47. Felix H, Brodelius P, Mosbach K (1981) Anal Biochem 116:462
48. Brodelius P, Nilsson K (1980) FEBS Lett 122:312
49. Nishida Y, Sato T, Tosa T, Chibata I (1979) Enzyme Microbial Tech 1:95
50. Sato T, Nishida Y, Tosa T, Chibata I (1979) Biochimica et Biophysica Acta 570:179
51. Takata I, Kayashina K, Tosa T, Chibata I (1982) J Ferment Technol 60:431
52. Takata I, Kayashina K, Tosa T, Chibata I (1982) J Appl Biochem 4:371
53. Wang HY, Hettwer DJ (1982) Biotechnol Bioeng 24:1827
54. Kuu WY, Polack JA (1983) Biotechnol Bioeng 25:1995
55. Takata I, Tosa T, Chibata I (1983) Appl Biochem Biotechnol 8:31
56. Wang HY, Lee SS, Takach Y, Cawthon L (1982) Biotechnol Bioeng Symp 12:139
57. Kierstan M, Buche C (1977) Biotechnol Bioeng 19:389
58. Hackel U, Klein J, Megnet R, Wagner F (1975) Eur J Appl Microbiol 1:291
59. Paul F, Vignais PM (1980) Enzyme Microbiol Technol 2:281
60. Cheetham PSJ (1979) Enzyme Microbiol Technol 1:183
61. Cheetham PSJ, Blunt KW, Bucke C (1979) Biotechnol Bioeng 21:2155
62. Veliky IA, Williams RE (1981) Biotechnol Lett 3:275
63. Birnbaum S, Pendleton R, Larsson PO, Mosbach K (1981) Biotechnol Lett 3:393
64. Jones A, Veliky IA (1981) Can J Bot 59:2095
65. Vieth WR, Wang SS, Saini R (1973) Biotechnol Bioeng 15:565
66. Venkatasubramanian K (1980) Desalination 35:353
67. Vieth WR, Venkatasubramamian K (1976) Methods Enzymol 44:243

68. Kolot FB (1981) Proc Biochem 16:2, 4, 9, 30
69. Venkatasubramanian K, Constantinides A, Vieth WR (1975) Enzyme Eng 3:29 (Pub 1978)
70. Yamada H, Yamada K, Kumagai H, Hino T, Okamura S (1978) Enzyme Eng 3:57
71. Hino T, Yamada H, Okamura S, Kojima H, Yokomoto Y Ito (1976) Japan Kokai 76,144,778, 13 December
72. Matsumoto K, Seijo H, Watanabe T, Karube I, Sahoh I, Suzuki S (1979) Anal Chim Acta 105:429
73. Karube I, Matsunaga T, Suzuki S (1980) Biotechnol Bioeng 22:2607
74. Tischer W, Tiemeyer W, Simon H (1980) Biochemie 62:331
75. Thonart P, Paquot M, Baijot B, Michaux M, Dervanne C (1982) Belgian Patent No BE890,811, 15 February
76. Parascandola P, Salvadore S, Scardi V (1982) J Ferment Technol 60:477
77. D'Souza SF, Kaul R, Nadkarni GB (1982) Biotechnol Bioeng 24:1701
78. de Rosa J, de Rosa M, Gambacorta A, Esposito E (1981) Biotechnol Bioeng 23:221
79. Updike SJ, Harris DR, Shrag E (1969) Nature 224:1122
80. Sommerville HJ, Mason JR, Ruffell RN (1977) Eur J Appl Microbiol 4:75
81. D'Souza SF, Nadkarni GB (1980) Biotechnol Bioeng 22:2191
82. Yamamoto K, Sato T, Tosa T, Chibata I (1974) Biotechnol Bioeng 16:1589
83. Koshchenko KA, Sukhodolskaya GV, Tyurin VS, Skryabin GK (1981) Eur J Appl Microbiol Biotechnol 12:161
84. Kokubu T, Karube I, Suzuki S (1978) Eur J Microbiol 5:233
85. Karube I, Dokuba T, Suzuki S (1981) Biotechnol Bioeng 23:29
86. Ohlson S, Larsson PO, Mosbach K (1978) Biotechnol Bioeng 20:1267
87. Slowinski W, Charm SE (1973) Biotechnol Bioeng 15:973
88. Morikawa Y, Karube I, Suzuki S (1980) Eur J Appl Microb Biotechnol 10:23
89. Morikawa Y, Karube I, Suzuki S (1979) Biotechnol Bioeng 21:261
90. Wheatley MA, Phillips CR (1981) Adv Biotechnol 2:47
91. Wheatley MA, Phillips CR (1983) Biotechnol Bioeng 25:623
92. Kumakura M, Yoshida M, Kaetsu I (1978) Eur J Appl Microbiol Biotechnol 6:13
93. Zueva NN, Shchrbakoba VN, Yakovleva VYa, Nikitia YuS, Avsyuk IV, Chan Tkhi Tuet Mai, Berezin IV (1980) Prikl Biokhim Mikrobiol 16:918
94. Stottmeister U (1979) Z Allg Mikrobiol 19:763
95. Berger R, Langhammer G (1980) Z Allg Mikrobiol 20(1):69
96. Uchida T, Watanabe T, Kato J, Chibata I (1978) Biotechnol Bioeng 20:255
97. Tanaka Y, Hayashi T, Kawashima K, Yokoyama T, Watenabe T (1982) Biotechnol Bioeng 24:857
98. Murata K, Kato J, Chibata I (1979) Biotechnol Bioeng 21:877
99. Godbole SS, D'Souza SF, Nadkarni GB (1983) Enzyme Microbiol Technol 5:125
100. Murata K, Tani K, Kato J, Chibata I (1981) Enzyme Microbiol Technol 3:233
101. Koshcheyenko KA, Turkina MV, Skryabin GK (1983) Enzyme Microbiol Technol 5:14
102. Morikawa Y, Karube I, Suzuki S (1980) Biotechnol Bioeng 22:1015
103. Freeman A, Aharonowitz Y (1981) Biotechnol Bioeng 23:2747
104. Tanaka A, Jin I-N, Kawamoto S, Fukui S (1979) Eur J Appl Microbiol Biotechnol 7:351
105. Fukui S, Sonomoto K, Itoh S, Tanaka A (1980) Biochemie 62:381
106. Fukui S, Tanaka A (1982) Ann Rev Microbiol 36:145
107. Sonomoto K, Usui W, Tanaka A, Fukui S (1983) Eur J Appl Microbiol Technol 17:203
108. Omata T, Atsuo T, Saburo F (1980) J Ferment Technol 58:339
109. Fukui S, Ahmed SA, Amata T, Tanaka A (1980) Eur J Appl Microbiol Biotechnol 10:289
110. Omata T, Tanaka A, Fukui S, Iida T (1979) Eur J Appl Microbiol Biotechnol 8:143
111. Omata T, Tanaka A, Yamane T, Fukui S (1979) Eur J Appl Microbiol Biotechnol 6:207
112. Fusee MC, Swann WE, Calton GJ (1981) Appl Environ Microbiol 42:672
113. Drioli E, Iorio G, Santoro R, DeRosa M, Gambacorta A, Nicolaus B (1982) J Mol Catal 14:247
114. Fukui S, Yokozaki K, Yamanaka S, Utagawa T, Takinami K, Hirose Y, Tanaka A, Sonomoto K (1982) Eur J Appl Microbiol Biotechnol 14:225
115. Fukui S, Tanaka A (1976) FEBS Lett 66:179

116. Tanaka A, Yasuhara S, Osumi M, Fukui S (1977) Eur J Biochem 80:193
117. Sonomoto K, Tanaka A, Omata T, Yamane T, Fukui S (1979) Eur J Appl Microbiol Biotechnol 6:325
118. Kumakura M, Yoshida M, Kactsu I (1979) Biotechnol Bioeng 21:679
119. Nippon Kayaku Co. Ltd. (1981) Japan Kokai Tokkyo Koho JP 81,140,889, 4 November
120. Bezman SA, Burtus PA, Izod TPJ, Thayer MA (1978) Photochem Photobiol 28:325
121. Japan Atomic Energy Research (1980) Japan Kokai Tokkyo Koho 8054,896, 22 April
122. Klein J, Wagner F (1980) Enzyme Eng 5:335
123. Klein J, Kressdorf B (1982) Biotechnol Lett 4:375
124. Shin Nenryoyu Kaihatsu Gijutsu Kenkyu Kumiai Japan Kokai Tokkyo Koho (1982) JP 57,198,088 [82,198,088], 4 December
125. Fukui S, Sada E, Tanaka A, Yamane T (1980) T Komata Jpn Kokai Tokkyo Koho 80 15,703, 4 February
126. Rouxhet PG, Van Haecht JL, Didelez J, Gerard P, Briquet M (1981) Enzyme Microb Technol 3:49
127. Sanraku-Ocean Co. Ltd. (1979) GB 1556584, 28 November
128. Arcuri EJ, Worden RM, Shumate SE (1980) Biotechnol Lett 2:499
129. Chibata I (ed) (1978) Immobilized enzymes research and development. Wiley, New York
130. Munir M (1982) Ger. Offen DE 3,038,219, 15 April
131. Novo Industri AS (1978) GB 1516704, 5 July
132. Hartmeier W (1981) Adv Biotechnol Proc Int Ferment Symp, 6th, 1980, 3:377
133. Hartmeier W (1981) Curr Dev Yeast Res Proc Int Yeast Symp 5th, 1980, p 105
134. Sivaraman H, Seetarama Rao B, Purdle AV, Sivaraman C (1982) Biotechnol Lett 4:359
135. Constantinides A (1980) Biotechnol Bioeng 22:119
136. Barbotin JN, Thomasset B (1980) Biochemie 62:359
137. Gulaya VE, Turkova J, Jirku V, Frydrychova A, Coupek J, Ananchenko SN (1979) Eur J Appl Microbiol Technol 8:43
138. Fletcher M (1979) In: Ellwood DC, Melling J, Rutter P (eds) Adhesion of microorganisms to surfaces. Academic Press, New York, p 87
139. Alemzadeh I, Maeda Y, Fazeli A (1977) J Ferment Technol 55:181
140. Lee CK, Long ME (1974) US Patent No 3,821,086, 28 June
141. Weeks MG, Munro PA, Spedding PL (1983) Biotechnol Bioeng 25:687
142. Weeks MG, Munro PA, Spedding PL (1981) Des Change, Proc Australas Chem Eng Conf 9th, p 493
143. Tsumura K, Kasumi T (1975) Japan Kokai 76,128,474, 28 April
144. Kokufuta E, Matsumoto W, Nakamura I (1982) Biotechnol Bioeng 24:1591
145. Hsiao HY, Chiang LC, Yang CM, Chen LF, Tsao GT (1983) Biotechnol Bioeng 25:363
146. Chibata I, Tosa T, Sato T (1979) Microbiol Technol 2:433
147. White FH, Portno AD (1978) J Inst Brew 84:228
148. Arcuri EJ (1982) Biotechnol Bioeng 24:595
149. Gerson DF, Zajic JE (1979) Immobilized microbial cells. ACS Symp Ser 106, p 29
150. Burns RG (1979) In: Ellwood DC, Melling J, Rutter P (eds) Adhesion of microorganisms to surfaces. Academic Press, New York, p 109
151. Kennedy JF (1979) Immobilized microbial cells. ACS Symp Ser 106, p 19
152. Kennedy JF, Humphreys JD, Barker A, Greenshields RN (1980) Enzyme Microbiol Technol 2:209
153. Atrat P, Hüller E, Hörhold C (1980) Z Allg Mikrobiol 20:79
154. Daugulis AJ, Brown NM, Cluett WR, Dunlop DB (1981) Biotechnol Lett 3:651
155. Setton OC, Gaddy JL (1980) Biotechnol Bioeng 22:1735
156. Rotman B (1960) Bactiological Rev 24:251
157. Tsumura N, Kasumi T (1977) US Patent No 4,001,082, 4 January
158. Fukushima M, Fujii T, Matsumoto K, Morishita M (1976) Japan Patent No 70884
159. Miyoshi T, Ishimatsu Y, Kimura S (1977) Japan Patent No 120185
160. Ghose TK, Chand S (1978) J Ferment Technol 56:315
161. Chand S, Ghose TK (1978) Bioconvers Cellul Subst Energy, Chem Microb Protein. Symp Proc [1st] 1977 (Pub 1978), p 573
162. Matteau PP, Saddler JN (1982) Biotechnol Lett 4:513

163. Matteau PP, Saddler JN (1982) Biotechnol Lett 4:715
164. Gudin C, Thomas D (1981) CR Seances Acad Sci Ser III Sci Vie 293:35
165. Ajinomoto Co., Inc. Japan Kokai Tokkyo Koho (1981) JP 57,144,989, 4 March
166. Karube I, Matsunaga T, Otomine Y, Suzuki S (1981) Enzyme Microbiol Technol 3:309
167. Matsunaga T, Karube I, Suzuki S (1978) Anal Chim Acta 99(2):233
168. Kayano H, Matsunaga T, Karube I, Suzuki S (1981) Biotechnol Bioeng 23:2283
169. Costamagna L (1981) Sci Tec Latt-Casearia 32:41
170. Kayano H, Karube I, Matsunaga T, Suzuki S, Nakayama O (1981) Eur J Appl Microbiol Biotechnol 12:1
171. Iwata Kagaku Kogyo Co, Ltd. Japan (1983) Kokai Tokkyo Koho JP 58 05,195 [83 05,195] (Cl. C12P7/48), 12 January
172. Tanabe Seiyaku Co, Ltd. Kewpie K.K. Japan Kokai Tokkyo Koho (1982) JP 82 18,986, 30 January
173. Mitsubishi Chemical Industries Co, Ltd. Japan Kokai Tokkyo Koho (1982) JP 82,110,192, 8 July
174. Yi Z-H, Rehm HJ (1982) Eur J Appl Microbiol Biotechnol 16:1
175. Chua JW, Eraeslan A, Kinoshita S, Taguchi H (1980) J Ferment Technol 58:123
176. Wada M, Kato J, Chibata I (1980) Eur J Appl Microbiol Biotechnol 10:275
177. Kim HS, Dewey DYRyu (1982) Biotechnol Bioeng 24:2167
178. Yamamoto K, Tosa T, Chibata I (1980) Biotechnol Bioeng 22:2045
179. Wada M, Uchida T, Kato J, Chibata I (1980) Biotechnol Bioeng 22:1175
180. Sawada H, Kinoshota S, Yoshida T, Taguchi H (1981) J Ferment Technol 59:111
181. Deo YM, Gaucher GM (1983) Biotechnol Lett 5:125
182. Frein EM, Montenecourt BS, Eveleigh DE (1982) Biotechnol Lett 4:287
183. Murata K, Tani K, Kato J, Chibata I (1980) Eur J Appl Microbiol Biotechnol 10:11
184. Linko Y-Y, Viskari R, Pohjola L, Linko P (1978) Enzyme Eng 4:345
185. DeBont JAM, Van Ginkel CG, Tramper J, Luyben KCAM (1983) Enzyme Microb Technol 5:55
186. Sawa Y, Kanayama K, Ochiai H (1980) Agric Biol Chem 44:1967
187. Alfermann AW, Schuller I, Reinhard E (1980) Plant Med 40:218
188. Ohlson S, Flygare S, Larsson PO, Mosbach K (1980) Eur J Appl Microbiol Biotechnol 10:1
189. Ohlson S, Larsson PO, Mosbach K (1979) Eur J Appl Microbiol Biotechnol 7:103
190. Kluge M, Klein J, Wagner F (1982) Biotechnol Lett 4:293
191. Veelken M, Pape H (1982) Eur J Appl Microbiol Biotechnol 15:206
192. Kurzatkowski W, Kurylowicz W, Paszkiewicz A (1982) Eur J Appl Microbiol Biotechnol 15:211
193. Suzuki S, Karube I (1979) Immobilized microbiological cells. ACS Symp Ser 106, p 59
194. Tramper J, Luyben KCAM, van den Tweel WJJ (1983) Eur J Appl Microbiol Biotechnol 17:13
195. Szwajcer E, Brodelius P, Mosbach K (1982) Enzyme Microbiol Technol 4:409
196. Brodelius P, Hagerdal B, Mosbach K (1980) Enzyme Eng 5:383
197. Tipayang P, Kozaki M (1982) J Ferment Technol 60:595
198. Spettoli P, Bottacin A, Nuti MP, Zamorani A (1982) Am J Enol Vitic 33:1
199. Stenroos SL, Linko Y-Y, Linko P (1982) Biotechnol Lett 4:159
200. Linko P, Stenroos S, Linko Y (1982) Biotechnol Lett 4:159
201. Navarro AR, Rubio MC, Callieri DAS (1983) Eur J Appl Microbiol Biotechnol 17:148
202. Shin Nenryoyu Kagaku Gijutsu Kenkyu Kumiai Japan Kokai Tokkyo Koho (1983) JP 58 13,391 [83, 13,391] (Cl. C12N11/10), 25 January; (1981) Appln 81/108,160, 13 July
203. Veliky IA, Williams RE (1981) Biotechnol Lett 3:275
204. Linko Y-Y, Jalanka H, Linko P (1981) Biotechnol Lett 3:263
205. Sliniger PJ, Bothast RJ, Black LT, McGhee JE (1982) Biotechnol Bioeng 24:2241
206. Pardonova B, Polednikova M, Sedova H, Kahler M, Ludvik J (1982) Brauwissenschaft 35:254
207. McGhee JE, St. Julian G, Detroy RW (1982) Appl Environ Microbiol 44:19
208. Williams D, Munnecke DM (1981) Biotechnol Bioeng 23:1813
209. Totsuka A, Hara S (1981) Hakkokogaku Kaishi 59:231

210. Margaritis A, Bajpai PK, Wallace JB (1981) Biotechnol Lett 3:613
211. Maleszka R, Veliky IA, Schneider H (1981) Biotechnol Lett 3:415
212. Linko Y-Y, Linko P (1981) Biotechnol Lett 3:21
213. McGhee JE, St Julian G, Detroy RW, Bothast RJ (1982) Biotechnol Bioeng 24:1155
214. Bayer N, Godtfredsen SE (1981) PCT Int Appl WO 81 03,339 (Cl. C12P7/08), 26 November; (1980) DK Appl 80/2,068, 12 May
215. Margaritis A, Bajpal P (1982) Biotechnol Bioeng 24(7):1483
216. Fukushima S (1980) Abstr 2nd German-Japanese Workshop on Enzyme Technology, p 38
217. Grote W, Lee KJ, Rogers PLL (1980) Biotechnol Lett 2:481
218. Haegerdal B, Mosbach K (1979) Food Proc Eng [Proc Int Congr], 2nd (Pub 1980) 2:129
219. Margaritis A, Wallace JB (1982) Biotechnol Bioeng Symp 12 (Symp Biotechnol Energy Prod Conserv 4th), p 147
220. Haeggstroem L (1980) Adv Biotechnol [Proc Int Ferment Symp], 6th 1980 (Pub 1981) 2:79
221. Foerberg C, Enfors SO, Haeggstroem L (1983) Eur J Appl Microbiol Biotechnol 17:143
222. Cho GH, Choi CY, Choi YD, Han MH (1981) Biotechnol Lett 3:667
223. Scherer P, Kluge PM, Klein J (1981) Biotechnol Bioeng 23:1057
224. Cheetham PSJ, Imber CE, Isherwood J (1982) Nature (London) 299:628
225. Bisping B, Rehm HJ (1982) Eur J Appl Microbiol Biotechnol 14:136
226. Nilsson I, Ohlson S (1982) Eur J Appl Microbiol Biotechnol 14:86
227. Hino T, Yamada H, Okamura S, Kojima H, Okamoto Y, Ito Y (1976) Japan Kokai 76,144,779 (Cl. C12K1/00), 13 December; (1975) Appl 75/69,005, 7 June
228. Karube I, Matsunaga T, Mitsuda S, Suzuki S (1977) Biotechnol Bioeng 19:1535
229. Marek M, Valentova O, Demnerova K, Jizba J, Blumaurerova M, Kas J (1981) Biotechnol Lett 3:327
230. Parascandola P, Scardi V (1982) Biotechnol Lett 4:753
231. Dhulster P, Parascandola P (1983) Enzyme Microbiol Technol 5:65
232. Kokubu T, Karube I, Suzuki S (1981) Biotechnol Bioeng 23:29
233. Mosbach K, Larsson PO (1970) Biotechnol Bioeng 12:19
234. Gulaya VE, Ananchenko SN, Torgov IV, Koshcheenko KA, Bychkova GG (1979) Bioorg Khim 5:768
235. Yang HS, Studebaker JF (1978) Biotechnol Bioeng 20:17
236. Kumakura M, Yoshida M, Kaetsu I (1978) Eur J Appl Microbiol Biotechnol 6:13
237. Atrat P, Hüller E, Hörhold C (1981) Eur J Appl Microbiol Biotechnol 12:157
238. Morikawa Y, Ochiai K, Karube I, Suzuki S (1979) Antimicrob Agents Chemother 15:126
239. Morikawa Y, Karube I, Suzuki S (1979) Biotechnol Bioeng 21:261
240. Pache W (1978) Eur J Appl Microbiol Biotechnol 5:171
241. Berezin IV, Yakovleva VI, Zueva NN, Malofeeva IV, Shcherbakova VN, Andreeva AP, Gubnitskij LI (1978) Dokl Akad Nauk SSSR 242:953
242. Zueva NN, Shcherbakova VN, Yakovleva VI, Avsyuk IV, Mai CT, Berezin IV (1980) Biokhimiya 45:2206
243. Zueva NN, Yakovleva VI, Avsyuk IV, Arens AK, Fechina VA, Berezin IV (1982) Prikl Biokhim Mikrobiol 18:681
244. Chibata I, Tosa T, Sato T (1974) Appl Microbiol 27:878
245. Tosa T, Sato T, Mori T, Chibata I (1974) Appl Microbiol 27:886
246. Yakovleva VI, Malofeeva IV, Zueva NN, Andreeva AP, Gubnitsky LS, Shcherbakova VN, Berezin IV (1979) Prikl Biokhim Mikrobiol 15:328
247. Bang WG, Lang S, Sahm H, Wagner F (1978) Preprint – Eur Congr Biotechnol, 1st, p 186 (DECHEMA: Frankfurt/Main)
248. Decottignies-Le Marechal P, Calderon-Seguin R, Vandecasteele JP, Azerad R (1979) Eur J Appl Microbiol Biotechnol 7:33
249. Azerad R, Calderon-Seguin R, Decottignies-Le Marechal P (1980) Bull Soc Chim Fr, Deuxieme Partie (1–2), 83
250. Bang WG, Behrendt U, Lang S, Wagner F (1983) Biotechnol Bioeng 25:1013
251. Sarkar JM, Mayaudon J (1983) Biotechnol Lett 5(3):201

252. Yamada H, Shimizu S, Shimada H, Tani Y, Takahashi S, Ohashi T (1980) Biochimie 62:395
253. Murata K, Tani K, Kato J, Chibata I (1978) Eur J Appl Microbiol Biotechnol 6:23
254. Murata K, Tani K, Kato J, Chibata I (1981) Eur J Appl Microbiol Biotechnol 11:72
255. Yamamoto K, Tosa T, Yamashita K, Chibata I (1976) Eur J Appl Microbiol 3:169
256. D'Souza SF, Nadkarni GB (1980) Biotechnol Bioeng 22:2179
257. Nadkarni GB, D'Souza SF (1981) Indian J Microbiol 21:244
258. Godbole SS, Kaul R, D'Souza SF, Nadkarni GB (1983) Biotechnol Bioeng 25:217
259. Iordan EP, Ikonnikov NP, Kovrizhnykh VA, Vorob'eva LI (1979) Prikl Biokhim Mikrobiol 15:515
260. Martin CKA, Perlmann D (1976) Biotechnol Bioeng 18:217
261. Ohmiya K, Ohashi H, Kobayashi T, Shimizu S (1977) Appl Environ Microbiol 33:137
262. Murata K, Uchida T, Tani K, Kato J, Chibata I (1979) Eur J Appl Microbiol Biotechnol 7:45
263. Saif SR, Tani Y, Ogata K (1975) J Ferment Technol 53:380
264. Mavituna F, Sinclair CG (1978) Preprint – Eur Congr Biotechnol, 1st, p 182 (DECHEMA: Frankfurt/Main)
265. Wheatley MA, Phillips CR (1980) Adv Biotechnol [Proc Int Ferment Symp], 6th (Pub 1981), 2:47
266. Pines G, Freeman A (1982) Eur J Appl Microbiol Biotechnol 16:75
267. Savino A, Lollini MN, Angeli G (1981) Ig Med 75:652
268. Nabe K, Izuo N, Yamdad S, Chibata I (1979) Appl Environ Microbiol 38:1056
269. Matsunaga T, Karube I, Suzuki S (1980) Biotechnol Bioeng 22:2607
270. Divies C (1977) Ann Microbiol 128B:349
271. Zueva NN, Yakovleva VI, Shcherbakova VN, Gubnitskij LS, Andreeva AP, Malofeeva IV, Berezin IV (1979) Biokhimiya 44:364
272. Dommergues YR, Diem HG, Divies C (1979) Appl Environ Microbiol 37:778
273. Divies C, Siess MH (1976) Ann Microbiol (Paris) 127B:525
274. Starostina NG, Lusta KA, Fikhte BA (1982) Prikl Biokhim Mikrobiol 18:225
275. Omata T, Tanaka A, Fukui S (1980) J Ferment Technol 58:339
276. Yokozeki K, Yamanaka S, Takinami K, Hirose Y, Tanaka A, Sonomoto K, Fukui S (1982) Eur J Appl Microbiol Biotechnol 14:1
277. Yongsmith B, Tanaka A, Fukui S (1980) Ann Rep ICME 3:263
278. Omata T, Iwamoto N, Kimura T, Tanaka A, Fukui S (1981) Eur J Appl Microbiol Biotechnol 11:199
279. Kimura A, Tatsutomi Y, Mizushima N, Tanaka A, Matsuno R, Fukuda H (1978) Eur J Appl Microbiol Biotechnol 5:13
280. Kumakura M, Kaetsu I (1983) Eur J Appl Microbiol Biotechnol 17:197
281. Tanaka A, Itoh N, Fukui S (1982) Agric Biol Chem 46:127
282. Hoq MM, Tanaka A, Fukui S (1981) Ann Rep ICME 4:139
283. Yamane T, Nakatani H, Sada E, Omata T, Tanaka A, Fukui S (1979) Biotechnol Bioeng 21:2133
284. Sonomoto K, Nomura K, Tanaka A, Fukui S (1982) Eur J Appl Microbiol Biotechnol 16:57
285. Fukui S, Tanaka A, Gellf G (1978) Enzyme Eng 4:299
286. Fukui S, Tanaka A (1980) Adv Biotechnol [Proc Int Ferment Symp], 6th (Pub 1981), 3:343
287. Sonomoto K, Hoq MM, Tanaka A, Fukui S (1983) Appl Environ Microbiol 45:436
288. Sonomoto K, Hoq MM, Tanaka A, Fukui S (1981) J Ferment Technol 59:465
289. Watanabe K, Itoh N, Tanaka A, Fukui S (1982) Agric Biol Chem 46:119
290. Egerer P, Simon H (1982) Biotechnol Lett 4:501
291. Japan Atomic Energy Research Institute (1980) Jpn Kokai Tokky Koho 80 54,896 (Cl. C12N11/04), 22 April; (1978) Appl 78/125,995, 13 October
292. Jirku V, Turkova J, Kuchynkova A, Krumphanzi V (1979) Eur J Appl Microbiol Biotechnol 6:217
293. Ziomek E, Martin WG, Veliky IA, Williams RE (1982) Enzyme Microbiol Technol 4:405

294. Turkova J, Gulaya VE, Jirku V, Ananchenko SN, Torgov IV, Frydrychova A, Coupek J (1982) Czech CS 202,959 (Cl. C12N11/04), 15 October; (1979) Appl 79/2,980, 28 April
295. Jirku V, Turkova J, Krumphanzl V (1980) Biotechnol Lett 2:59
296. Jirku V, Turkova J, Veruovic B, Kubanek V (1980) Biotechnol Lett 2:451
297. Schnarr GW, Szarek WA (1977) Appl Environ Microbiol 33:732
298. Felix HR, Mosbach K (1982) Biotechnol Lett 4(3):181
299. Novo Industri AS GB 1516704 P5.7.78, A26.8.75. PRUS 28.8.74 (501292)
300. Jirku V, Macek T, Vanek T, Krumphanzl V, Kubanek V (1981) Biotechnol Lett 3:447
301. Munir M (1982) Ger Offen DE 3,038,219 (Cl. C12P19/12). 15 April; (1980) Appl 9 October
302. Sivaraman H, Seetarama Rao B, Pundle AV, Sivaraman C (1982) Biotechnol Lett 4:359
303. Jack TR (1977) Biotechnol Bioeng 19:631
304. Ramesh V, Singh C (1981) Enzyme Microbiol Technol 3:246
305. Navarro JM, Durand G (1981) Ann Microbiol 132B(2):241
306. Lee CK, Long ME (1974) U.S. 3,821,086 (Cl. 195/116; C12b), 28 June; (1971) Appl 161,337, 9 July
307. Nazly N, Knowles CJ (1981) Biotechnol Lett 3:363
308. Kennedy JF (1978) Enzyme Eng 4:323
309. Kennedy JF (1979) Immobilized microbial cells, ACS Symp Ser 106, p 119
310. Heinrich M, Rehm HJ (1982) Eur J Appl Microbiol Biotechnol 15:88
311. Elliot ET, Cole CV, Fairbanks BC, Woods LE, Bryant RJ, Coleman DC (1983) Soil Biol Biochem 15:85
312. Navarro JM, Durand G (1977) Eur J Appl Microbiol 4:243
313. Marcipar A, Cochet N, Brackenridge L, Lebeault JM (1979) Biotech Lett 1:65
314. Arinbasarova AYu, Koshcheenko KA (1980) Prikl Biokhim Mikrobiol 16:854
315. Lambert GR, Daday A, Smith GD (1979) FEBS Lett 101:125
316. Messing RA, Oppermann RA, Ramsey WS, Takeguchi MM (1981) U.S. 4,246,349 (Cl. 435-176, C12N11/14), 20 January; (1978) Appl 939,176, 5 September
317. Ghose TK, Tyagi RD (1982) J Mol Catal 16:11
318. De Bremaeker M, Gennen M, Kayem GJ, Rouxhet PG, Van Haecht JL (1980) Belg 884,877 (Cl. C12N), 16 December; (1980) Appl 201,826, 22 August
319. Horne PN, Hsu HW (1983) Anal Biochem 129:72
320. Ghommidh C, Navarro JM, Durand G (1981) Biotechnol Lett 3:93
321. Moo-Young M, Lamptey J, Robinson CW (1980) Biotechnol Lett 2:541
322. Navarro AR, Lucca ME, Callieri DAS (1982) Acta Cient Venez 33:214
323. Ryu YW, Navarro JM, Durand G (1982) Eur J Appl Microbiol Biotechnol 15:1
324. Leung K-L, Joshim S, Yamazaki H (1983) Enzyme Microbiol Technol 5:181
325. Vossough M, Laroche M, Navarro JM, Faup G, Leprince A (1982) Water Res 16:995
326. Hilali A, Molina JAE (1979) Appl Environ Microbiol 38:1140
327. Seyhan E, Kirwan DJ (1979) Biotechnol Bioeng 21:271
328. DeNicola K, Kirwan DJ (1980) Biotechnol Bioeng 22:1283
329. Atrat P, Groh H (1981) Z Allg Mikrobiol 21:3
330. Ash SG (1979) In: Ellwood DC, Melling J, Rutter P (eds) Adhesion of microorganisms to surfaces. Academic Press, New York, p 73

3 Special Problems and Extended Applications

Although immobilization is normally applied to fermentations involving enzymes or microbial cells, there are many related areas in which the same principle is used. These areas, which are outside of the present scope, include tissue culture, immunology, dialysis, genetic engineering and affinity chromatography. In this chapter, special problems and techniques encountered in immobilization work are discussed, together with extensions beyond microbial cells and to multiple enzyme systems.

3.1 Special Problems and Techniques

3.1.1 Inhibition of Enzyme Activity

Enzyme inhibition can be understood by means of the active sites model in which the substrate is a substance containing a limited number of active sites onto which the molecules of enzyme orient themselves according to their molecular conformations to form a reactive complex which then goes on to form the product. Inhibition may be competitive or non-competitive. In competitive inhibition, the presence of a second substance other than the substrate competes with substrate molecules for active sites, but does not interfere with the adsorption between substrate and enzyme other than by a steric hindrance effect. An increase of substrate concentration in the case of an enzyme reaction showing competitive inhibition results in an increase in the original reaction rate. In non-competitive inhibition, the inhibiting agent not only occupies an active site on the substrate but also interferes with the adsorption of enzyme molecules to the nearby sites on the substrate so that an increase of substrate concentration does not result in an increase in enzyme activity. This distinction may not be clearcut if, in an enzyme reaction, an intermediate is produced which shows competitive inhibition at low substrate concentration, but exhibits enhancement and finally inhibition of activity as product concentration increases [1].

The inhibiting action of enzymes may be derived from the substrate, the products of a fermentation, or occasionally the carrier material [2]. In a microbial fermentation, inhibition effects in addition to the above can be expected on the growth of the microorganism. An advantage of a system utilizing immobilized cells is that continuous fermentation may allow control of substrate concentration, nutrient level and product removal so as to reduce inhibition from the substrate or the products. For example, inhibition from the substrates can be reduced

Table 3.1. Inhibition of growth or enzyme activity in immobilized cells

Microbe	Carrier	Inhibition agent	Ref.
Saccharomyces cerevisiae	Ca alginate gel	Substrate and product	[86]
Saccharomyces cerevisiae, *Saccharomyces uvarum,* *Zymomonas mobilis*	Ca alginate gel	Substrate	[87]
Saccharomyces cerevisiae	Ca alginate gel	Product	[88]
Saccharomyces cerevisiae		Substrate	[89]
Gluconobacter oxydans	Ca alginate	Product	[90]
Alcaligenes faecalis	Polyacrylamide	Product	[91]
Schizosaccharomyces pombe 007 and *Leuconostoc* *mesenteroides IAM 1233*	Calcium alginate gels		[92]
Corynebacterium simplex	Collagen tanned with glutaraldehyde	Product	[1]
Escherichia coli	Polyacrylamide gel	Substrate	[93]
Nocardia erythropolis	DEAE-cellulose and Al alginates and polyacrylates	Specific	[94]
Escherichia coli	Polyacrylamide gel	Substrate	[95]
Yeast microbes from *Kloeckera sp 2201*	Photo-crosslinkable resins	Specific	[96]
Escherichia coli	Polyacrylamide		[97]
Rhizopus nigricans	Polyacrylamide, alginate and agar gels		[98]
Acetobacter aceti	Cordierite	Product	[99]

by using a controlled rate of addition of substrate, and toxic products can be removed as soon as they are formed either in a continuous flow system or by dialysis.

Inhibition of β-glucosidase by glucose produced in the hydrolysis of cellobiose can be reduced by co-immobilizing a yeast which consumes the glucose as soon as it is formed [3]. On the other hand, inhibiting effects from substrates are exploited in the separation by affinity chromatography of closely similar enzymes that nonetheless show different adsorption patterns on the substrates because of inhibition.

A compilation of recent examples of inhibition in immobilized cell fermentation is provided in Table 3.1.

3.1.2 Insolubility and Limited Diffusion

Since most of the enzymes encountered in fermentation with immobilized cells are intracellular, the cell wall poses a barrier to diffusion of nutrient and sterically bulky substrates into and products out of the cell. In addition to diffusion, many microorganisms can utilize a wall turn-over mechanism to release enzymes to the environment and for purposes such as interacting with other cells via cell attachment or binding of pathogens by immunological processes [4]. The exocellular

Table 3.2. Improvement of permeability in immobilized cells by addition of reagents

Cell/carrier	Product	Reagent added	Action	Ref.
Protoplasts of *Brevibacterium flavum*	Glutamic acid	Penicillin	Antibiotic	[100]
Arthrobacter simplex photo-cross-linkable resin pre-polymers	Hydrocortisone Δ' dehydrogenation	Phenazine metho-sulphate or 2,6-dichlorophenol-indophenol		[18]
Nocardia rhodocrous photo-crosslink-able prepolymer	Androst-1,4-diene-3,17 dione			[22]
Pseudomonas testos-teroni polyacryl-amide gel	Δ'-dehydrogena-tion of Reich-stein's substance S	Phenazine metho-sulphate	Electron acceptor	[101]
Arthrobacter simplex polybutadiene	Steroid dehydrogenation	Surfactants		[102]
Escherichia coli polyacrylamide	L-Tryptophan	Nonionic detergents		[97]
Brevibacterium am-moniagenes poly-acrylamide gel	NADP	Detergents	Activation	[103]
Escherichia coli polyacrylamide	L-Tryptophan	Indole, excess pyruvate	Substrate inhibitor	[95]
Clostridium butyricum agar	Hydrogen	Riboflavin	Enhances hydrogen production	[104]

synthesis of steroid glycoalkaloids by the plant cells of *Solarum aviculare* has been documented [5].

Hydrophobic substrates pose some difficulties because of their low solubility in the aqueous system, which is generally the system of choice in fermentations because of the ease of maintaining a suitable physiological environment. Steroids, and in some cases, coenzymes such as NADPH, may cause problems because of low solubility or low diffusibility through the cell wall [6].

The permeability of immobilized microbial cells can be increased by three methods, namely, thermal [7–9], addition of reagent to modify the cell wall, and use of aqueous/organic solvent mixtures (Table 3.2). There is a concurrent effect of some of these reagents on the carrier porosity, as discussed separately. It has been demonstrated [7–9] that the condition of the carrier can determine the degree of permeabilization of the immobilized cells. The use of gluteraldehyde as a cross-linking agent on *Saccharomyces carlsbergensis* bound on porous glass was found to influence the overall cell metabolism, and favour the product ethanol over carbon dioxide [10].

There is an enhancement action of surfactant and detergents on the rate of reaction in immobilized cells, especially in bacterial leaching [11]. Although the exact mechanisms are so far unknown, it is suspected [11] that improved gas transfer is involved in addition to more extensive wetting. Detergents and emulsifiers such as bile extract, bile acid, and deoxycholic acid were screened for their action in suppressing succinic acid formation as a byproduct in the transformation of fumaric acid to malic acid by *Brevibacterium ammoniagenes* [12]. It has been found that bile extract has the greatest effect in by-product suppression and that detergents are not effective. The solubilizing characteristics of bile extracts from bound enzymes suggest that the action of the surfactant is to overcome permeability barriers in the cells so that substrates and products can diffuse more readily. Furthermore, such treatment produces operational stability in the immobilized cells relative to the untreated intact cells. Similar stabilizing effects on the enzymes of immobilized plant cells upon permeabilization treatment have been noted [13]. The antibiotic nystatin causes spores to form on the cell wall [6]; penicillin probably acts through the cell wall also.

Another method of inducing permeability in cell walls is to use mixed organic solvent systems [14–16] in a fermentation, sometimes in conjunction with a carrier system which allows modifications of polarity (Table 3.3). The latter approach, developed by Fukui et al. [17] for water-insoluble substrates, employs a photo-crosslinkable polyurethane prepolymer system in which the two ends of a poly-alkyleneglycol backbone have been connected to hydroxyalkylacrylate moieties with bifunctional diisocyanates. Structural parameters that can be varied include the chain length of the polyglycol and the nature of the alkylene groups in the polyglycol. Thus, for example, if polypropyleneglycol is used instead of polyethyleneglycol, a more lipophilic carrier is obtained, which produces a higher hydrocortisone to prednisolone transformation ratio [18]. Similarly, polyurethane car-

Table 3.3. Use of solvent mixtures in immobilized cell fermentation

Cell/carrier	Product	Solvent system	Effect	Ref.
Several kinds of enzymes and microbial cells carrageenan		Water-miscible organic solvents	Operation stability	[105]
Catharanthus roseus agarose beads polyurethane matrix		Dimethyl-sulphoxide	Permeabilization	[106]
Clostridium acetobutylicum	Acetone and BuOH	Polyethylene glycol 8000	2-Phase systems	[107]
Caldariella acidophila cellulose acetate membranes		Acetone	Increased initial activity	[108]
		Toluene	Permeabilization	[7]
		Organic solvent	Activation	[103]
Enterobacter aerogenes photo-crosslinkable resin prepolymer, ENT-400, urethane prepolymer		Dimethyl-sulphoxide	Increased operational stability, substrate loading	[109]
Escherichia coli 18-crown-6 ether		18 Crown ether	Enhanced mycelial growth	[110]
		Butanol	Increased activity	[111]

riers of the diisocyanate/polyalkyleneglycol/diisocyanate structure have been developed to permit a certain control of lipophilicity. These carriers of controlled lipophilicity are often used to advantage in conjunction with a solvent system in which the polarity can be adjusted. Generally such an environment may prove to be drastically different from that of aqueous systems so that the activity of free enzymes decreases as a result of failure to maintain the requisite molecular conformation. In this sense, immobilization of the enzyme on a carrier or immobilization of viable or even dead microbial cells confers stability to the enzyme.

In reviewing the use of organic solvent systems, FUKUI and TANAKA [14] have included homogeneous systems such as two-phase water/water miscible organic solvent systems as well as organic solvent systems. In most fermentations homogeneous systems are desirable because they adapt easily to continuous production. Typical solvents used include dimethylsulphoxide, methanol, butanol, acetone, chloroform, polyethyleneglycol, ether, toluene, benzene, n-heptane with lipophilic substrates such as cholesterol [19] and dehydroepiandrosterone [20]. It has been established that the rate of enzyme reaction of cells immobilized on a gel is influenced by the distribution of the substrate between the solution and the lipophilic carrier gel.

A more non-polar (lipophilic) solvent system evidently causes more substrate to enter the carrier gel, resulting in a higher reaction rate. By choosing a carrier/solvent system of the required polarity, it is possible to direct the course of some steroid transformations [21, 22]. Thus testosterone (non-polar) can be oxidized to 1,4-androstene-3,17-diene when catalyzed by the enzyme 17β-hydroxysteroid dehydrogenase or to Δ^1-dehydrotestosterone by the electron acceptor phenazine methosulphate (polar). The former reaction is favoured by using cells immobilized on a non-polar carrier and the latter reaction by a polar carrier in which the polar phenazine methosulfate can penetrate, to produce Δ^1-dehydrotestosterone which inhibits the enzyme 17β-hydroxysteroid dehydrogenase.

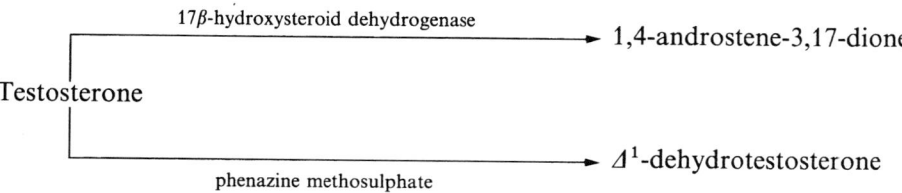

Yeast cells permeabilized by heat or acetone treatment and immobilized on photo-crosslinkable or polyurethane prepolymers have been used to phosphorylate AMP or ADP to ATP [23–25].

3.2 Extension of Immobilized Materials

By far the most frequent materials used in immobilization are single enzymes and single whole microbial cells from the classes: bacteria; algae, yeast and moulds; and the green algae. However, under various conditions, other biological materials have been immobilized, including whole cells from plants and animals [26], various organelles from plant, animal or microbial cells, and various materials co-immobilized together on the same carrier.

Electron micrographs of animal and plant sub-cellular materials such as red blood cells, spheroplasts and chromatophores have been immobilized on protein foam crosslinked with glutaraldehyde and serum albumin and their morphology related to their observed kinetic activities [27].

Perhaps the extension areas undergoing the most rapid advancement and having the highest potential for industrial application concern the immobilization of plant cells as well as organelles such as chloroplasts and chromatophores isolated from plant or bacterial sources.

3.2.1 Plant Cells

The great variety of alkaloids, polysaccharides and proteins produced by plant cells in vivo serves as an incentive to develop a high yield, continuous production process for these materials. Plant cells from, for example, *Digitalis* [28], *Daucus* [29, 30], *Catharanthus* [28], *Morinda* [28], and *Vicia* [31], *Cannabis* [32], *Ipomoea* [32], and *Solanum* [5], may be isolated and used in freely suspended or immobilized cell cultures. Fermentation by immobilized plant cells is reviewed by BRODELIUS and MOSBACH [13] and BIRNBAUM et al. [33].

Whole plant cells are most often immobilized in polysaccharides such as alginate, agar, agarose and carrageenan, all of which have the advantage of maintaining the viability of the immobilized plant cells [28]. Various permeabilizing procedures are usually applied to facilitate diffusion of substrates and products into and out of the cell, and the fermentation controlled in a continuous state by substrate addition and product removal. Permeabilization can be effected by the action of solvents such as dimethylsulphoxide [34–38].

Cells of *Catharanthus roseus* treated in this way retained their activity in the enzymes hexokinase and glucose-5-phosphate dehydrogenase longer than untreated cells. The higher yield of ajmalicine alkaloids from cathenamine resulted from the ease of diffusion through the plasma membrane as well as the co-immobilization of the coenzyme NADPH. Other permeabilizing agents investigated by FELIX et al. [38] based on analogous permeabilizing action on microbial cells include toluene [34, 39], ether [38], antibiotics [40–42], proteins [38] such as poly-L-lysin, cytochrome C, protamine sulfate and lipophilic agents [38] such as lysolecithin and nystatin.

Biosynthetic reaction schemes of different complexity have been effected by immobilized plant cell systems, ranging from the functional group modification very often encountered in steroid transformations, to the assembly of molecular precursors, to, when the problem of by-product contamination has been reduced by the availability of high producing cells [43, 44], total synthesis from simple starting materials.

Examples of algae immobilization, in particular their role as oxygen suppliers to fermentation reactions, have been reported [45].

3.2.2 Animal Cells

Only a limited amount of work has been conducted so far on animal cell immobilization, an account of which has been provided by BIRNBAUM et al. [46]. Animal cells immobilized include human skin fibroblasts, human kidney carcinoma, and rat colon carcinoma erythro-leukemic cells [26].

Chitosan, agarose and alginic acid have been used as carriers. Gelatin has been found to be the best carrier for cells of human skin fibroblasts and rat colon carcinoma. Techniques for testing cell growth and for harvesting were developed with possible application to analysis and steroid transformation. Another example of animal cell immobilization is the dialysis membrane electrode utilizing a thin slice of porcine liver for the detection of glutamine [47]. Glutaminase activity has been found to be enhanced and prolonged, giving excellent sensitivity and selectivity. Red blood cells immobilized on protein foam have been examined by the scanning electron microscope [27].

3.2.3 Subcellular Materials

Immobilization of subcellular materials (Table 3.4) such as chloroplasts, chromatophores, mitochondria, and microbodies is reviewed by OCHIAI et al. [48]. Chloroplasts can be isolated by homogenization [49] and centrifuging [50]. Mitochondria are isolated from yeast by treatment with bacterial lytic enzyme to release the protoplasts, which after homogenizing, are centrifuged to give a mitochondria fraction [51]. Yeast (Kloeckera sp. no. 2201) microbodies are isolated by an analogous method using fractionation by differential centrifugation and sucrose density-gradient ultracentrifugation [52].

Chloroplasts. Immobilization of chloroplasts has received major effort and represents an extreme form of permeabilization of a plant cell because in this case the cell chloroplasts are enclosed only by the membrane and the resulting rapid diffusion of material through this membrane provides excellent conditions for maintaining continuous processes.

Chloroplasts contain enzymes that play important roles in the light-limited stage of photosynthesis containing the two electron flow systems (PSII and PSI) as well as in the dark stage of photosynthesis, catalyzing reactions such as phosphorylation of ADP to ATP, reduction of cofactor NADP to NADPH, removal of hydrogen from water, and fixation of CO_2. However, PSI and especially PSII are not stable in the chloroplasts in vitro although immobilization is known to provide a certain degree of stability for investigation. Among the chloroplast immobilization carriers, including most carriers employed in routine cell-immobilization work, stabilization effects on PSII and PSI have been reported for polyvinylalcohol [53], DEAE-cellulose [54] and calcium alginate [55]. The four photoreactions of the light-limited step have an obvious significance in cell immobilization reactions. The reactions involving regeneration of the cofactors NADPH and ATP can be coupled to other immobilized cells for enzymes in a fermentation. If the process were economical, the hydrogen generated from the PSII could be utilized as a fuel or in a fuel cell.

Table 3.4. Immobilized organelles and sub-cellular components

Material immobilized	Carrier	Product/function	Ref.
Spinach chloroplast	Glass forming monomers	Study of PS II in photosynthesis	[112]
Chloroplast	Polyvinyl alcohol	Study of PS II in photosynthesis	[53]
Chloroplast	Polyacrylamide gel	Photosynthesis study	[113]
Chloroplast	Microcapsules		[114]
Algal filaments	Glass beads	H_2, O_2 production	[115]
Spinach chloroplast	Polyacrylamide gel	Photosynthesis study	[50]
Chloroplasts	DEAE cellulose	Photosynthesis study	[54]
Protoplasts of *Vicia faba*	Calcium alginate	Morphological study	[31]
Rat liver mitochondria	Alkylsilated glass	Studies on respiration	[56]
Yeast mitochondria	Photo-crosslinkable resin prepolymer or polyurethane prepolymer	Preparation of adenylate kinase	[51]
Thylakoids, spheroplasts, chromatophores, red blood cells	Protein foam	Morphological observation with electron microscopy	[27]
Yeast peroximes *Kloeckera* sp.	Hydrophilio methane prepolymers Pu-9	Preparation of adenylate kinase	[59]
Yeast microbodies (peroxizones)	Photo-crosslinkable prepolymer ENT-4000	Enzyme studies	[116]
Yeast microbodies *Kloecke* sp. 2201	Photo-crosslinkable prepolymers		[96]
Microbodies (peroxisomes) of *Kloeckera*	Photo-crosslinkable prepolymer ENT or albumin/ glutaraldehyde	Hydrocortisone to prednisolone trans-formation	[58]
Bacterial chromatophores from *Rhodopseudomonas capsulata*	Various carriers	ATP production study	[117]
Bacterial chromatophores from *Rhodospirillum rubrum*	Polyacrylamide gel	Photo-phosphorylation study	[118]
Spores of *Clostridium acetobutylicum*	Calcium alginate gel	Study of acetone and butanol formation	[65]
Conidia of *Penicillium urticae*	κ-Carrageenan beads	Antibiotic patulin	[63]
Spores of *Clostridium butyricum*	Calcium alginate	*n*-Butanol, iso-butanol production	[64]
Spores of *Tusarium* sp.		6-Amino penicillamic acid	[62]
Spores of *Curvularia lunata*	Calcium alginate	Steroid 11 β-hydroxylation	[60]
Spores of *Curvularia lunata*	Photo-crosslinkable resin prepolymer	Hydrocortisone	[61]
Spores of *Aspergillus* sp. and *Penicillium* sp.		Sucrose inversion	[119]

Mitochondria. Mitochondria are organelles conducting the functions of cell metabolism, electrical transmission, and maintaining osmotic work and cell motility by a complex series of enzyme reactions in the Krebs Cycle. They oxidize amino acids, fatty acids and pyruvic acid derived from simple sugars producing energy as ATP by oxidative phosphorylation of ADP. Because their enzyme system is highly complex, mitochondria generally do not maintain their activity in vitro. Immobilization apparently affords some stability, allowing for investigations to be performed into the maintenance of a continuous interchange of material through the mitochondrial membrane.

Respiration in an environment approximating dynamic in vivo conditions has been studied using a rat liver mitochondria monolayer immobilized on alkylsilated glass to which continuous replenishment of culture medium is possible [56].

Immobilized mitochondria have also been used for study of the activities of catalase, flavin-dependent alcohol oxidase and adenylate kinase [51], which catalyzes the phosphorylation of ADP to ATP.

Microbodies (peroxisomes). Yeast microbodies (peroxisomes) are subcellular organelles consisting of a single membrane enclosing spherical particles containing a series of enzymes which produce hydrogen peroxide and convert it to oxygen and water. Yeast microbodies containing the enzymes catalase, alcohol oxidase and D-amino acid oxidase have been immobilized on photocrosslinkable resin prepolymers [57]. Activities have been found to be lower than those in free microbodies. It has later been found that when a mixture of polyethyleneglycolhydroxyethyleneacrylate polymers and poly(propylene glycol)-hydroxyethylacrylate polymers, or albumin is used as carrier, the catalase, D-amino acid oxidase and oxidase activities are similar to those of free peroxisomes. This immobilized system in the photocrosslinkable prepolymer mixture has been utilized for transformation of the steroid hydrocortisone to prednisolone [58]. Immobilized peroxisomes have been used in the preparation of adenylate kinase [59].

Spores. The advantages of spores are that they are simple to immobilize, relatively easy to isolate from the cells and are able to withstand environments that are detrimental to the corresponding cells. Spores are usually immobilized in a carrier and then allowed to germinate to develop the mycelial structure in a uniform immobilized system that has better mechanical properties and operational stability than obtainable by direct mycelial immobilization. Calcium alginate [60] has been preferred as a carrier over polyacrylamide for spores of *Curvularia lunata,* although the resulting steroid hydroxylation activity is still unstable. *Curvularia lunata* spores have also been successfully immobilized in photocrosslinkable resin prepolymers and urethane prepolymers [61]. In situ germination of these spores produced a system of immobilized mycelia difficult to obtain by standard immobilization methods and possessing improved stability with respect to the activities of the steroid 11-β-hydroxylation.

A certain enzyme activity is sometimes associated with spores, for example, the hydrolytic enzyme in *Fusarium sp.* spores [62]. However, enzyme activity is enhanced after the spores are induced to germinate, usually by incubating in a nutrient. The procedure of obtaining immobilized cells by germination of immobilized spores in a carrier allows the handling of microorganisms which have fragile structures which otherwise make it difficult to obtain a homogeneous, stable, leak-proof immobilized system of mycelial material [60, 63–65].

3.2.4 Multiple Enzyme Systems

Co-Immobilization. Co-immobilization is the process of binding on the same carrier two or more biological materials among which some biochemical interaction occurs in a fermentation. Examples are given in Table 3.5. Technically, enzymes, whole microbial cells and organelles can be co-immobilized in any combination, consistent with their biochemical compatibility, function and the objective of the fermentation. Co-immobilization of enzymes is the out-growth of enzyme immobilization techniques in which two or more enzymes or cofactors are in close proximity to effect the desired reaction. The cofactors most often needed are the adenine nucleotide pairs ATP/ADP, $NAD^+/NADP^+$, and CoA. Co-immobilization of enzymes on a molecular level has been successful, mainly in the study of multiple enzyme system in nature such as the glucose-ethanol transformation (9 enzymes) [66], the urea cycle (4 enzymes) [67], and the model mitochondria system (3 enzymes) [68].

Some enzyme-enzyme co-immobilization systems of industrial importance are described by BUCKE and WISEMAN [69]. These are the glucose oxidase/catalase system [70], the glucose isomerase/glucoamylase system [71] and the glucose isomerase/lactose system [72].

Because of its fundamental importance, multiple enzyme immobilization at the molecular level has recently been reviewed [6, 73, 74]. However, as a means for the production of fermentation products, multiple enzyme systems have some serious disadvantages, chiefly in the cost of pure enzymes and cofactors, the difficulty of extracting, purifying or regenerating the enzymes, and the possibility of product contamination due to leakage of the immobilized enzymes. Furthermore, the presence of multiple enzymes and cofactors may impose so many, often incompatible, constraints on the physical conditions of the system that it becomes exceedingly difficult to operate within the limits demanded by the activity and stability of the enzymes and cofactors.

Since most of the immobilized enzymes originate from whole cells, it becomes desirable to immobilize these cells on carriers, and, for those cells whose function in fermentation involves multiple enzymes, to immobilize these in the living state so that the multiple enzymes can operate in their original physiological environment. Living cell immobilization has since become an important area of investigation and application and has recently been the subject of excellent comprehensive reviews [75]. An aspect of living cell immobilization having relevance in co-immobilization is production by the living cells of co-enzymes and other primary metabolites that can be utilized in a separate fermentation in a different cell or enzyme system. Many such co-immobilization systems are at present being studied. Problems include loss of permeability of the cells, by-product formation from enzymes not connected with the fermentation scheme, and maintaining the course

Table 3.5. Co-immobilization with living whole cells

	Carrier	Description	Ref.
I. Enzyme/cell co-immobilization			
Pepsin/*Saccharomyces cerevisiae*	Glutaraldehyde/tannin	Wine making	[120]
β-Glucosidase/baker's yeast + lactose	Calcium alginate	Ethanol from cellobiose or lactose	[121]
Amyloglucosidase/ brewer's yeast	Calcium alginate	Wort fermentation	[78]
Glucoamylate/yeast		Beer, wort fermentation	[122]
β-Glucosidase/baker's yeast		Ethanol from cellobiose	[3]
Cellulase/baker's yeast	Glutaraldehyde/tannin crosslinking	Ethanol from cellobiose	[123]
β-Glucosidase/*Saccharomyces cerevisiae*	Calcium alginate	Ethanol from cellobiose and cellulose hydrolytic products	[79]
Δ'-Dehydrogenase/ *Curvularia lunata*		Predonisolone from 11-deoxycortisone	[124]
Escherichia coli glutathione synthetases, acetate kinase, dextran-bound ATP, and ATP analogs	Polyacrylamide gel	Glutathione synthesis	[81]
II. Cell/cell co-immobilization			
Escherichia coli (glutathione synthetases) *Saccharomyces cerevisiae* (glycolytic pathway)	Carrageenan gel	Glutathione synthesis with coupled ATP regeneration	[81]
Providencia sp. PCM 1298 bacteria) *Chlorella vulgaris* (algae, O$_2$ supplier)	Agarose	(Keto acid production) through amino acid oxidase	[45]
Gluconobacter oxydans (bacteria) *Chlorella pyrenoidosa* (algae)	Calcium alginate	Transforming or glycerol to dihydroxyacetone	[85]
Rhodospirillum rubrum/ *Klebsiella pneumoniae*	Agar	Hydrogen	[83]
Porcine kidney tissue/ *Sarcina flava*		Glutamine to ammonia	[82]
Gluconobacter melanogenes/ *Pseudomonas syringus*		2-Keto-L-gluconic acid from L-sorbose	[80]
Saccharomyces cerevisiae/ *Brevibacterium ammoniagenes*	Ethyl cellulose	Coenzyme regeneration	[24]
Saccharomyces cerevisiae/ *Brevibacterium ammoniagenes*	Cellulose acetate/ butylate	NADP production	[125]

of cell growth towards production of the required cofactors instead of towards increase of biomass. Cellulose has been co-immobilized with yeast which has the ability to hydrolyse cellubiose in the production of ethanol [76]. In the brewing industry, enzymes such as pepsin [77] (anti-foaming agent) and amyloglucosidase [78] (low-calorie beer production) are used.

Enzyme-Cell Co-Immobilization. In this variation, enzymes are co-immobilized with microbial cells to augment the microbial enzyme system which may be deficient in certain enzymes. The microorganism most often used for this purpose is *Saccharomyces cervisiae,* which, though itself a complex system of enzymes and cofactors, cannot effectively complete some fermentations, notably that of transforming cellulose all the way to ethanol. The reason for this failure to complete the fermentation is that in the series of degradation stages of cellulose, intermediate products such as cellubiose, glucose and even the end product, ethanol act as product inhibitors, so that a mixture of cellubiose and glucose usually results. This situation may be alleviated by continuous product removal by membrane dialysis together with co-immobilization of β-glucosidase to hydrolyse cellubiose to glucose [79].

Cell-Cell Co-Immobilization. The most important type of cell/cell co-immobilization is that in which one cell produces a cofactor necessary for another cell. Thus *Saccharomyces cerevisiae, Escherichia coli,* and *Brevibacterium ammoniagenes* have often been used as generators of ATP to provide energy for co-immobilized cells. Co-immobilized cells have been surveyed by CHIBATA et al. [74]. An early example of co-immobilization of whole cells led to the production of 2-keto-L-galonic acid [80]. Co-immobilization of two different kinds of whole cells has been adopted by MURATA et al. [81] in connection with glutathione synthesis, which involves the group of glutathione synthetases found in *Escherichia coli* cells and energized by ATP generated in the acetate kinase system, also occurring in these cells. In order to facilitate the transferring and build-up of ATP inside the whole cells, dextran-bound ATP (or its analogs) have been co-immobilized with the microbial cells in a reaction which gave an increased production rate of glutathione. Since whole cells of *Saccharomyces cerevisiae* are equipped with the enzyme system for glutathione as well as the ATP regeneration system via the glycolytic pathway, they have been co-immobilized with *Escherichia coli* cells to augment the supply of ATP for glutathione synthesis. The method of co-enzyme transfer has also been applied to the co-immobilized system of *Saccharomyces cerevisiae* (S-adenosyl-L-methionine production) and *B. ammoniagenes* (NADP production). Both co-immobilized systems prove to be superior to single cell systems. Another early example of whole cell co-immobilization involves porcine kidney tissues and the bacterium *Sarcina flava* to detect glutamine by converting it to ammonia [82].

In the production of hydrogen from cellulose hydrolysate by the cells of *Rhodospirillum rubrum,* WEETALL et al. [83] have found that the presence of *Klebsiella pneumoniae,* initially a contaminant and subsequently added as co-immobilized cells, is responsible for increased efficiency and increased hydrogen production compared to single cells, including *Clostridium butyricum* [84]. The postulated explanation is that the co-immobilized cells of *Klebsiella pneumoniae* produced, in addition to a small amount of hydrogen, organic acids and carbohydrates from dextrose, thus facilitating the break-down of dextrose to hydrogen gas by *Rhodospirillum rubrum* cells.

Another mode of operation in co-immobilization systems involves the transfer of oxygen produced by the alga *Chlorella vulgaris* to the bacterial cells of *Providencia* sp. to increase the production of keto acid by amino acid oxidase action under red-light illumination [45]. Independently, ADLERCREUTZ et al. [85] found that in situ supply of oxygen by the co-immobilized algae *Chlorella pyrenidosa* increased the yield of the glycerol to dihydroxyacetone in the fermentation of *Gluconobacter oxydans*.

Finally, *Pseudomonas syringus,* a source of L-sorbosone oxidase, has been co-immobilized with *Gluconobacter melanogenes* for the conversion of L-sorbose to 2-keto-L-gulonic acid via L-sorbosone. The enzyme from the co-immobilized cells catalyses the second step of the conversion [80]. These examples suggest that transfer of such diverse materials as ATP, NADP, oxygen, organo acids, carbohydrates, enzymes and co-enzymes among co-immobilized whole cells of bacterial, algal, and animal origin is possible with a notable exception [80] and generally leads to improved yields of products.

ADLERCREUTZ et al. [85] have suggested the following desiderata for fermentations using complementary co-immobilized whole cells: full capacity utilization of the cell metabolic pathway rather than increase of biomass, high cell density and choice of cell couples that are complementary in their metabolic function.

3.3 Immobilized Cells in Industrial Production

In industrial fermentation, next to the traditional techniques based on free microbial whole cells, the most often used techniques include immobilized enzymes and immobilized cells. Free enzymes are less frequently used because of their high cost, the need for recycling via extraction and purification, and the possibility of product contamination by leakage. The industrial uses of immobilized cells have been the topic of many reviews.

References

1. Constantinides A (1980) Biotechnol Bioeng 22:119
2. Jeanfils J, Collard F (1983) Eur J Appl Microbiol Biotechnol 17(4):254
3. Hägerdal B, Mosbach K (1980) Fund Proc Eng 2:129
4. Ellwood DC, Melling J, Rutter P (1979) In: Ellwood DC, Melling J, Rutter P (eds) Adhesion of microorganisms to surfaces. Academic Press, New York, p 3
5. Jirku V, Macek T, Vanek T, Krumphanzl V, Kubanek V (1981) Biotechnol Lett 3:447
6. Mosbach K (1983) Phil Trans R Soc Lond B 300:355
7. Van Keulen MA, Vellenga K, Joosten GEH (1981) Biotechnol Bioeng 23:1437
8. Kan JK, Shuler ML (1978) AIChE Symp Ser 74:31
9. Jirku V, Kurkova J, Kuchynkova A, Krumphanzl V (1979) Eur J Appl Microbiol Biotechnol 6:217
10. Navarro JM, Durand G (1977) Eur J Appl Microbiol 4:243

11. Ash SG (1979) In: Ellwood DC, Melling J, Rutter P (eds) Adhesion of microorganisms to surfaces. Academic Press, New York, p 62
12. Yamamoto K, Tosa T, Yamashita K, Chibata I (1976) Eur J Appl Microbiol 3:169
13. Brodelius P, Mosbach K (1982) Adv Appl Microbiol 28:1
14. Fukui S, Tanaka A (1982) Ann Rev Microbiol 36:145
15. Weetall HH, Vann WP (1976) Biotechnol Bioeng 18:105
16. Khibanov AM, Samokhim GP, Martinek K, Berezin IV (1977) Biotechnol Bioeng 19:1351
17. Fukui S, Sonomoto K, Itoh S, Tanaka A (1980) Biochemie 62:381
18. Sonomoto K, Tanaka A, Omata T, Yamane T, Fukui S (1979) Eur J Appl Microbiol Biotechnol 6:325
19. Omata T, Iida T, Tanaka A, Fukui S (1979) Eur J Appl Microbiol 8:143
20. Omata T, Tanaka A, Fukui S (1980) J Ferment Technol 58:339
21. Fukui S, Ahmed SA, Omata T, Tanaka A (1980) Eur J Appl Microbiol 10:289
22. Yamane T, Nakatani H, Sada E, OmataT, Tanaka A, Fukui S (1979) Biotechnol Bioeng 21:2133
23. Asada M, Morimoto K, Nakanishi K, Matsuno R, Tanaka A, Kimura A, Kamikubo T (1979) Agric Biol Chem 43:1773
24. Samejima H, Kimura K, Ado Y, Suzuki Y, Tadokoro T (1978) Enzyme Eng 4:237
25. Kimura A, Tatsutomi Y, Mizushima N, Tanaka A, Matsuno R, Fukuda H (1978) Eur J Appl Microbiol 5:13
26. Nilsson K, Mosbach K (1980) FEBS Lett 118:145
27. Barbotin JN, Thomasset B (1980) Biochemie (Paris) 62:359
28. Brodelius P, Deus B, Mosbach K, Zank MH (1979) FEBS Lett 103:93
29. Jones A, Veliky IA (1981) Eur J Microbiol Biotechnol 13:84
30. Veliky IA, Jones A (1981) Biotechnol Lett 3:511
31. Scheurich P, Schnabel H, Zimmermann V, Klein J (1980) Biochim Biophys Acta 598:654
32. Jones A, Veliky IA (1981) Can J Bot 59(11):2095
33. Birnbaum S, Larsson PO, Mosbach K (1983) In: Scouten WH (ed) Solid phase biochemistry – Analytical and synthetic aspects. Wiley, New York, p 742
34. Delmer DP, Mills SE (1969) Plant Physiol 44:153
35. Delmer DP (1979) Plant Physiol 64:623
36. Milura GA, Mills SE (1971) Plant Physiol 47:483
37. Fritsch G, Griesebach H (1975) Phytochemistry 14:2437
38. Felix H, Brodelius P, Mosbach K (1981) Anal Biochem 116:462
39. Lerner HR, Ben-Bassad D, Reinhold L, Poljakoff-Mayber A (1978) Plant Physiol 61:213
40. Mudd JB, Kleinschmidt MG (1970) Plant Physiol 45:517
41. Gottlieb D, Carter HE, Wu L, Sloneker JH (1960) Phytophathology 50:594
42. Gottlieb D, Carter HE, Sloneker JH, Wu LC, Gaudy E (1961) Phytophathology 51:321
43. Zenk MH, El-Shagai H, Schulte U (1975) Planta Med Suppl 79
44. Zenk MH, El-Shagai H, Arens H, Stöchigt J, Weiler EW, Deuw B (1977) In: Barz W, Reinhard E, Zenk MH (eds) Plant tissue culture and its biotechnological application. Springer, Berlin Heidelberg New York, p 27
45. Wikstrom P, Szwajcer E, Brodelius P, Nilsson K, Mosbach K (1982) Biotech Lett 4:153
46. Birnbaum S, Larsson P-O, Mosbach K (1983) In: Scouten WH (ed) Solid phase biochemistry – Analytical and synthetic aspects. Wiley, New York, p 745
47. Rechnitz GA, Arnold MA, Meyerhoff ME (1979) Nature 278:466
48. Ochiai H, Tanaka A, Fukui S (1983) Appl Biochem Bioeng 4:153
49. Shioi Y, Tamai H, Sasa T (1978) Plant Cell Physiol 19:203
50. Ochiai H, Shibata H, Matsuo T, Kashinokuchi K, Yakawa M (1977) Agric Biol Chem 41:721
51. Tanaka A, Hazi N, Gellf G, Fukui S (1980) Agric Biol Chem 44:2399
52. Fukui S, Kawamoto S, Yasuhara S, Tanaka A (1975) Eur J Biochem 59:561
53. Ochiai H, Shibata H, Matsuo T, Hashinokuchi K, Inamura I (1978) Agric Biol Chem 42:683
54. Shioi Y, Sasa T (1979) FEBS Lett 101:311
55. Gisby PE, Hall DO (1980) Nature (London) 287:251

88

56. Arkles B, Brinigar WS (1975) J Biol Chem 250:8856
57. Tanaka A, Yasuhara S, Gellf S, Osumi M, Fukui S (1978) Eur J Appl Microbiol 5:1227
58. Fukui S, Tanaka A, Gellf G (1978) Enzyme Eng 4:299
59. Tanaka A, Kawamoto S, Fukui S (1979) Eur J Apl Microbiol 7:351
60. Ohlson S, Flygare S, Larsson P-O, Mosbach K (1980) Eur J Appl Microbiol 10:1
61. Sonomoto K, Hoq MM, Tanaka A, Fukui S (1981) J Ferment Technol 59:465
62. Charles M (1980) Abstr 6th Int Ferment Symp, p 120
63. Deo YM, Gancher GM (1982) Biotechnol Lett 5:125
64. Krouwell PG, van den Laan MM, Kossen NWF (1980) Biotechnol Lett 2:253
65. Haggstrom L, Molin N (1980) Biotechnology Lett 2:241
66. Mosbach K, Mattiason B (1978) Current topics in cellular regulation, vol 14. Academic Press, New York, p 197
67. Okamoto H, Tipayang P, Inanda Y (1980) Biochim Biophys Acta 611:35
68. Srere PA, Mattiasson B, Mosbach K (1973) Proc Natl Acad Sci USA 70:1534
69. Bucke C, Wiseman A (1981) Chem Ind 4:234
70. Hartmeier W, Tegge G (1979) Starch 31:348
71. Hollo J, Laszlo E, Hoschke A (1975) Starch 27:232
72. Weetall HH, Detar CC (1974) Biotechnol Bioeng 16:1537
73. Kolot FB (1980) Proc Biochem 15:2
74. Chibata I, Tosa T, Sato T (1983) Adv Biotechnol Proc 1:203
75. Kennedy JF, Cabral JMS (1983) Appl Biochem Bioeng 4:189
76. Hartmeier W (1981) Adv Biotechnol 3:377
77. Hartmeier W (1981) Curr Dev Yeast Res 105
78. Godtfredsen SE, Ottesen M, Svensson B (1981) Proc Congr Eur Brew Conv 18th, p 505
79. Hagerdal B, Lopez Leiva M, Mattiasson B (1980) Desalination 35:365
80. Martin CKA, Perlman D (1976) Eur J Appl Microbiol 3:91
81. Murata K, Tani K, Kato J, Chibata I (1980) Biochemie 62:347
82. Mascini M, Rechnitz GA (1980) Anal Chim Acta 116:169
83. Weetall HH, Sharma BP, Detar CC (1981) Biotech Bioeng 23:605
84. Karube I, Matsunaya T, Tsuru S, Suzuki S (1976) Biochem Biophys Acta 444:338
85. Adlercreutz P, Holst O, Mattiasson B (1982) Enzyme Microbiol Technol 4:395
86. Lee EH, Ahn JC, Ryu DDY (1983) Enzyme Microbiol Technol 5:41
87. McGhee JE, St Julian G, Detroy RW (1982) Biotechnol Bioeng 24:1155
88. Williams D, Munnecke DM (1981) Biotechnol Bioeng 23:1813
89. Wada M, Kato J, Chibata I (1981) Eur J Appl Microbiol Biotechnol 11:67
90. Holst O, Enfors S-O, Mattiasson R (1982) Eur J Appl Microbiol Biotechnol 14:64
91. Wheatley MA, Phillips CR (1980) Adv Biotechnol [Proc Int Ferment Symp] 6th (Pub 1981), 2:47 (1981)
92. Totsuka A, Hara S (1981) Hakkokagakukaishi 59:231
93. Azerad R, Calderon-Sequin R, Decottignies-Le Marechal P (1980) Bull Soc Chim Fr, Deuxieme Partie (1–2):83
94. Atrat P, Huller E, Horhold C, Buchar MJ, Arinibasarova AY, Koschtschejenko KA (1980) Z Allg Mikrobiol 20:159
95. Decottignies-Le Marechal P, Calderon-Seguin R, Vandecasteele JP, Azerad R (1979) J Appl Microbiol Biotechnol 7:33
96. Tanaka A, Yashuhara S, Osmui M, Fukui S (1977) Eur J Biochem 80:193
97. Bang WG, Lang S (1978) Preprint – Eur Cong Biotechnol 1st, p 186
98. Maddox IS, Dunnill P, Lilly MD (1981) Biotechnol Bioeng 23:345 .
99. Ghommidh C, Navarro JM, Durand G (1981) Biotechnol Lett 3:93
100. Ajinomoto Co, Inc Japan Kokai Tokkyo (1982) JP 57,144,989 [82,144,989] (Cl. C12P13/04), 7 September
101. Yang HS, Studebaker JF (1978) Biotechnol Bioeng 20:17–25
102. Fukui S, Sada E, Tanaka A, Yamane T, Komata T (1980) Japan Kokai Tokkyo Koho 80 15,703 (Cl. C12N11/08), 4 February
103. Murata K, Kato J, Chibata I (1979) Biotechnol Bioeng 21:877
104. Karube I, Urano N, Matsunaga T, Suzuki S (1982) Eur J Appl Microbiol Biotechnol 16:5

105. Tosa T, Sato T, Mori T, Yamamoto K, Takata I, Nishida Y, Chibata I (1979) Biotechnol Bioeng 21:1697
106. Felix HR, Mosbach K (1982) Biotechnol Lett 4:181
107. Mattiasson B, Suominen M, Andersson E, Haggstrom L, Albertsson PA, Hahn-Hagerdal B (1982) Enzyme Eng 6:153
108. De Rosa M, Gambacorta A, Esposito E, Drioli E, Gaeta S (1980) Biochimie 62:517
109. Fukui S, Yokozeki K, Yamanata S, Utagawa T, Takinami K, Hirose Y, Tanaka A, Sonomoto K (1982) Eur J Appl Microbiol Biotechnol 14:225
110. Tso WW (1980) Biotechnol Lett 2:519
111. Ohlson S, Larsson PO, Mosbach K (1978) Biotechnol Bioeng 20:1267
112. Yoshii F, Fujimura T, Kaetsu I (1981) Biotechnol Bioeng 23:833
113. Karube I (1979) Biotechnol Bioeng 21:253
114. Kitajima M, Butler WL (1976) Plant Physiol 57:746
115. Lambert GR, Daday A, Smith GD (1979) FEBS Lett 101:125
116. Tanaka A, Yashuara S, Gellf G, Osumi M, Fukui S (1978) Eur J Appl Microbiol Biotechnol 5:17
117. Garde VL, Thomasset B, Tanaka A, Gellf G, Thomas D (1981) Eur J Appl Microbiol Biotechnol 11:133
118. Yang HS, Leung KH, Archer MC (1976) Biotechnol Bioeng 18:1425
119. Johnson DE, Ciegler A (1969) Arch Biochem Biophys 130:384
120. Hartmeier W (1980) Curr Dev Yeast Res [Proc Int Yeast Symp] 5th (Pub 1981), p 105 (1981)
121. Haegerdal B (1980) Acta Chem Scand Ser B B34:611
122. Hartmeier W, Muecke I (1982) Util Enzymes Technol Aliment Symp Int 519
123. Hartmeier W (1980) Adv Biotechnol [Proc Int Ferment Symp] 6th (Pub 1981), 3:377 (1981)
124. Mosbach K, Larsson PO (1970) Biotechnol Bioeng 12:19
125. Ado Y, Kimura K, Samejima H (1980) Enzyme Eng 5:295

4 Properties of Immobilized Cell System

4.1 Properties of the Microbial Cells

4.1.1 Cells Used in Immobilization

In general, the range of microbial cells used in immobilization approximately co-incides with that used in fermentation processes, and reflects the desire to improve economically important processes by attaching the microorganisms to solid supports. Of the class of bacteria (*Schizomycetes*), the three orders of *Pseudomonadales*, *Eubacteriales* and *Actinomycetalis* are most frequently encountered and represented by the genera *Pseudomonas*, *Acetobacter*, *Azotobacter*, *Rhizobium*, *Streptococcus*, *Lactobacillus*, *Corynebacterium*, *Escherichia*, *Bacillus*, *Clostridium*, *Nocardia*, *Actinomyces*, and *Streptomyces* [1].

Other important microorganisms are the algae, yeast and molds (Division Thallophyta), comprising the classes of Phycomycetes, and Ascomycetes, the most frequently encountered genera being *Rhizopus*, *Mucor*, *Saccharomyces*, *Penicillium* and *Aspergillus*. Among the class of blue green algae (the Schizophyceae), some species are occasionally used in immobilization work.

Table 4.1 lists the genera and species of some of the microbial cells immobilized in recent literature. In addition to these, plant cells and cell organelles are employed with increasing frequency in immobilization systems.

Table 4.1. Cells common in immobilization work

Acetobacter	*Actinoplanes*	*Bacillus*
sp.	missouriensis	subtilis
suboxydans		licheniformis
xylinum	*Alcaligenes*	coagulans
aceti	faecalis	sp.
	entrophus	megaterium
Acetobularia		amyloliquefaciens
mediterranea	*Arthrobacter*	pseudomonas
	simplex	
Achromobacter	globiformis	*Botrycoccus*
aceris		braunii
butyri	*Azotobacter*	*Brevibacterium* (i.s.)[a]
liquidium	vinelandii	ammoniagenes (i.s.)
	chroöcoccum	fulvum (i.s.)
		(lysed cells of
Actinomyces	*Bacterium*	lactofermentum (i.s.))
roseochromogenes	cadaveris	fuscum (i.s.)

Table 4.1 (continued)

Caldariella
  acidophila

Citrobacter
  freundii

Clostridium
  acetobutyricum
  acidiurici
  spec. LA 1
  butyricum
  thermocellum
  beijerinckii

Corynebacterium
  glutamicum
  simplex
  dismutans
  sp.

Curvularia
  lunata

Desulfovibrio
  desulfuricans

Enterobacter
  aerogenes

Erwinia
  herbicola
  rhapontici

Escherichia
  coli
  freundii

Gluconobacter
  melanogenes IFO 3293
  oxydans

Haemophilus
  influenzae

Halicystis
  parvula

Hansenula
  polymorpha
  jadinii
  anomala

Klebsiella
  pneumoniae

Kloeckera
  sp. 2201

Kluyveri
  citrophila

Khryveromyces
  lactis
  fragilis
  marxianus

Lacterobacillus
  casei
  bulgarisus
  delbrueckii
  lactis
  sp.
  thermophilus
  vaccinnostercus
  arabinosus

Lens
  culinaris
  culinaris lectin

Leuconostoc
  mesenteriodes
  oenos ML 34

Litomosoides
  carinii

Mastigocladus
  laminosus

Mastomys
  natalensis

Methanomonas
  sp.

Methanosarcina
  barkeri

Methylomonas
  rubra
  flegellata

Micrococcus
  luteus

Mortierella
  vinacea

Mycobacterium
  rubrum
  globiforme
  phlei
  smegmatis
  pyl

Mycococcus
  mucosum

Neissenia
  gonorrhoeae

Nitrobacter
  sp.

Nitrosomonas
  europaea

Norcadia
  (lysed cells)
  rhodocrous
  erythropolis
  opaca
  fructifera (i.s.)

Pachysolen
  tannophilus

Pediococcus
  cerevisiae

Phasedus
  vulgaris

Photobacterium
  phosphoreum

Porphilidium
  cruentum

Proprionibacterium
  shermanii
  technicum
  arabinosum
  freundanreichii

Protaminobacter
  rubrum

Proteus
  vulgaris
  settgeri
  mirabilis
  morganii

Providencia
  sp.

Pseudomonas
  sp.
  polycolor
  putida
  testosteroni
  (lysed cells)
  fluorescens
  aeruginosa
  dacuhae (i.s.)
  solanacearum
  denitrificans (i.s.)
  paucimolilis

Red usnea

Rhizobium
  japonicum
  nigricans

Table 4.1 (continued)

Rhodopseudomonas capsulata	*Tricoderma* reesei E-58 sp.	*Rhizopus* stolonifer nigricans
Rhodospirillum rubrum	*Trigonopsis* variabilis	*Penicillium* cyaneofulvum spiculisporum vitale chrysogenum ulticae urticae
Rhodotorula sp. mucilaginosa minuta	*Trichosporon* brassicae cutaneum	
Ruminococcus albus	*Valonia* atricularis	*Neurospora* crassa
Saccharomycopsis lipolytica	*Vibrio* cholerae alginolyticus parahaemolyticus vulnificus anguillarum	*Fusarium* moniliforme oxysporum lini sp.
Salmonella typhimurium		
Sarcina flava ureae	*Xanthomonas* vesicatoria	*Chlorella* sp. pyrenoidosa vulgaris regularis
Serratia marcescens	*Zymomonas* mobilis	
Sporobolomyces sp.	*Candida* boidinii tropicalis lipolytica fumicola	*Aspergillus* ochraceus niger oryzae awramosi phoenicis
Sporosarcina ureae		
Sternphylium loti	*Zygosaccharomyces* lactis	*Anacystis* nidulans
Streptococcus faecium lactis pyogenes	*Schizosaccharomyces* pombe	*Anabaena* cylindrica sp.
	Scenedesmus obliquus	*Vicia* faba
Streptomyces sp. phaeochromogenes clavuligerus fradiae aureofaciens viridochromogenes libani tendae Tue 901 griseus	*Mucor*	*Solanum* av
	Saccharomyces carlsbergensis cerevisiae (whole cell or mitochondria) uvarum serratia paradoxus fragilis bayanus anamensis formosensis pastorianus	*Morinda* citrifolia
		Medicago littoralis
		Lichens
Torulosis candida lipofera		*Digitalis* lanata
		Daucus carota
Treponema pallidum		*Catharanthus* roseus

[a] i.s. = nomina incerta sedis.

4.1.2 The Growth Cycle

The growth of microbial cells immobilized on a solid carrier essentially follows the pattern of free cells. Extensive reviews on this topic are available (for example [2]).

After a given medium has been innoculated with a cell culture, growth of the cells, expressed as the logarithm of the number of cells, passes through a number of stages in the growth cycle. There are four characteristic stages on this curve: the lag phase, the exponential growth phase, the stationary phase and the exponential death phase. In addition to these, there are transitional phases: an accelerating phase just before the exponential phase, and the decelerating phase after the exponential phase (Fig. 4.1). During the lag and accelerating phases, the microorganisms are adapting in preparation for growth. During these phases growth is evident in the size of cells, that is, in an increase in protoplasm, enzyme membrane and other building materials rather than in the population. The rate of reproduction is generally low so that these phases are characterized by long generation times, although nutrient uptake may be rapid.

In the exponential phase, cell growth is derived from metabolism of substrates to produce primary metabolites, material necessary for the rapid increase of biomass. Growth in this phase shows a linear relationship on the log microbial population versus time curve. In this phase, growth as expressed by the rate of increase in the number of microorganisms bears a constant relationship to that expressed in terms of mass. By applying cyclical perturbations to the growing cells, for example, by changing the temperature of the supply of an essential nutrient which has been deliberately withheld, the cells can be made to grow synchronously [3, 4] so that most of the cells in the culture undergo the same stage in cell division at the same time.

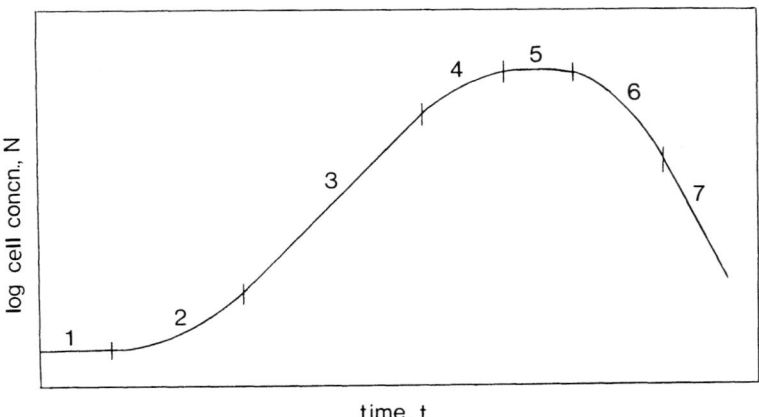

Fig. 4.1. Phases of growth in batch culture of microorganisms. During the logarithmic phase, most of the cells are viable; after this the number of viable cells is significantly less than the number of cells. *1* Lag phase. *2* Acceleration phase. *3* Logarithmic growth phase. *4* Retardation phase. *5* Stationary phase. *6* Accelerating death phase. *7* Logarithmic death phase

If one assumes that the cells in the exponential growth phase reproduce by binary fission, that is,

$$B = b \cdot 2^n,$$

where

  b = number of cells at the beginning
  B = number of cells at the end of the exponential period
  n = number of generations,

then, by taking logarithms of each side,

$$\log B = \log b + n \log 2$$

or

$$\frac{\log B - \log b}{\log 2} = n.$$

The duration of one generation or the generation time, g, is defined as:

$$g = \frac{t_e}{n},$$

where t_e is the total time of the exponential growth period. Therefore,

$$g = \frac{t_e \log 2}{\log B - \log b}.$$

The growth rate constant, K, is defined as:

$$K = \frac{n}{t_e}$$

$$= \frac{\log B - \log b}{t_e \log 2}$$

or

$$K = \frac{2.303 \, (\log B - \log b)}{t_e}.$$

During microbial growth, nutrients may be depleted and toxic waste products may accumulate, thereby slowing the growth rate as well as increasing the overall death rate, ultimately giving rise to a net deceleration in the growth rate. However, in a continuous fermentation, where nutrient supply and product and waste removal are constantly provided, the exponential growth period may be prolonged considerably. In an unperturbed environment, however, the exponential growth phase will eventually slow down and reach a stationary phase in which the microbial population remains constant. In this phase, production of secondary metabolites becomes important. These products are not strictly required in biomass production, but nonetheless may have great commercial value. The energy required by the cells to operate in the stationary phase is much less than that required for exponential growth [5] so that operational stability of cells in this phase can be maintained for a prolonged period of time. In a continuous fermentation, this stationary phase can be induced in microorganisms undergoing exponential growth by controlling the flow rate of one particular nutrient so that it is present at a limiting concentration, while all other nutrients are present in excess in the medium.

By choosing a glucose-based nutrient that will support product formation but not cell growth, the fermentation reaction of *Clostridium acetobutylium* may be diverted to the production of acetone and butanol instead of biomass [6]. Furthermore, enzyme activity may be maintained by a pulsewise addition of the nutrient. These methods are useful in directing the energy of the cells to produce desirable fermentation products instead of simply increasing their biomass. This objective can also be achieved by addition of an antibiotic, for example novabiocin [7], to suppress cell division and thus biomass production.

When the death rate overtakes the growth rate, the cell population is in the exponential death phase. Cells in this stage are usually deformed or lysed; however, if the requisite enzyme is retained, they can still undergo fermentation. Reactivation of cells in this stage may be possible by providing the necessary nutrient and growing conditions.

In general, relative to free cells, whole cells immobilized on a carrier in most cases show an increased enzyme activity with a longer half-life and a longer storage time. The enzyme of the cells is more stable towards heat and radiation [8]. Attachment to a carrier often affords protection against unfavorable environmental effects; immobilized cells generally show better operational stability, for example, a higher rate of growth, shorter generation time [3, 9], increased productivity, and a longer period of growth or viability. Immobilized cells can also survive longer storage times and can be reactivated [10] more readily on being supplied with a suitable nutrient. Immobilized cells tend to retain their shape and rigidity [11], although under conditions of rapid growth the size of immobilized cells is generally smaller than that of free cells. Attachment on a carrier usually also provides some protection against cell lysis due to unfavorable conditions. Availability of substrate and disposal of waste are not normally hindered by surface immobilization [3]. After prolonged immobilization and repeated use on a granular carrier in fermentation, cell lysis has been found to be more serious in the centre of the carrier granule than at the surface, where cells remain viable for longer periods [12]. It may be rationalized that in the centre of the granules, poor transport of nutrients and wastes by diffusion renders conditions less suitable for cell survival.

4.1.3 Viability

Criteria for cell viability [13, 14] include microscopic appearance, cell population, specific uptake of oxygen, specific evolution of carbon dioxide, consumption of substrate, and capability of reculturing [15, 16]. However, no single criterion, nor any combination of them, can be considered adequate. It is also important to note that viability should not be compared with signs of enzymatic activity, which can persist even though the cells are dead. For example, oxygen consumption may be due to the enzyme oxidase rather than the respiration of living cells. Moreover, evidence of enzyme activity may be modified by product or substrate inhibition [17, 18]. This situation is also true for other physiological actions of the cells. Success in re-incubating a culture or an apparent population increase does not necessarily mean that viability exists, because live cells may be produced from the unattached fraction. For carriers such as agar that can be dissolved totally to release all of their attached microorganisms, WAGNER and KLEIN [19] have suggested plate counting as the best means of establishing viability. The concept of cell viability is closely tied to that of endogenous metabolism or maintenance energy, a concise review of which is given by LAMANNA et al. [20]. Endogenous metabolism is the physiologic activity of viable cells maintained in the resting state. It may not include cell growth but often includes secondary metabolite production [21, 22].

4.1.4 Electron Microscopy

Direct investigation of the process of immobilization is possible by microscopic observation, using either light microscopy or electron microscopy, which can be divided into scanning electron microscopy (SEM) and transmission electron microscopy (TEM). In SEM, the specimen to be observed is usually fixed by a cross-linking agent, e.g., glutaraldehyde together with a staining agent, then dehydrated either by washing in mixtures containing increasing concentrations of ethanol in water, finally in absolute ethanol, then freeze-dried, or dried by the critical-point method. The specimen is placed in a high vacuum (10^{-5} Torr) and coated with a thin layer of gold to make the surface more efficient in electron scattering. For observation, a beam of electrons is directed onto the specimen and the resulting back-scattered electrons and secondary electrons collected, focussed, amplified and either displayed on a phosphorescent screen or recorded on a photographic film. The image obtained by SEM is a picture of the surface of the sample with a very great depth of field (generally about ½ of the breadth of field) so that an object can be viewed in perspective. The perspective effect can be controlled by choosing the angle between the incident electron beam and the detector to intercept a different portion of the backscattered electrons from the sample. The resultant change in shadow size brings out the surface topography of the object. Alternatively, all the electrons from the sample can be collected at the detector to produce an image that is evenly illuminated, without any shadows. Resolution in the Angstrom range is possible under exceptionally favorable conditions, but generally resolution of about 100 Å is routine. Internal structures of a sample can be

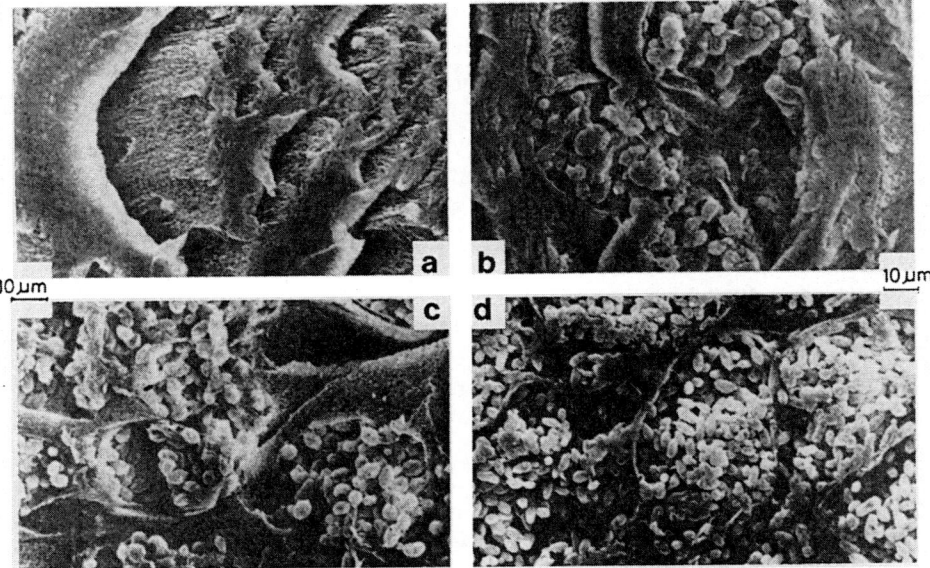

Fig. 4.2a–d. Changes in cell population in gel beads with incubation in the complete medium [25]. **a** Immediately after entrapment of cells; **b** after 16 h incubation (initial phase of logarithmic growth); **c** after 40 h incubation (final phase of logarithmic growth); **d** after 1200 h incubation (stationary growth phase)

observed by SEM at the fracture surface. In the transmission mode, sample preparation involves fixing with various reagents, including those containing metal salts, to impart electron density to the sample. The sample is then dehydrated, embedded and cut into sections of a few hundred angstroms. Observation involves passing an electron beam through the sample and collecting the transmitted electrons at a detector. Observations under the SEM are used often in the applications described below.

Identification of Microorganisms and Enzymes. Microorganism identification is an obvious application of microscopy. Identification can be made of the desired strain to be used in a fermentation as well as of contaminating strains or mutants. Electron micrographs of cell sections have been used in determining the localization of enzymes in prokaryotic cells [23]. In the more common technique, reaction product deposition, the enzyme in its active form is treated with a substrate giving a product which is or can be made electron dense for electron microscopic observation. In the less common technique, localization by immunocytochemistry, specific antigens are used to bind the enzymes. Coupling of the resulting complex with reagents such as ferritin or peroxidase provide a way of bonding electron-dense reagents to the site for electron microscopic observation. Cellular enzymes may occur on the cell surface, in the cell wall, or in or around the cell plasma, these locations being determined by examining the thin sections of the localized cell enzymes. By means of electron microscopy, the enzymes catalase, D-amino acid oxidase and a flavin-dependent alcohol oxidase have been localized in the micro-

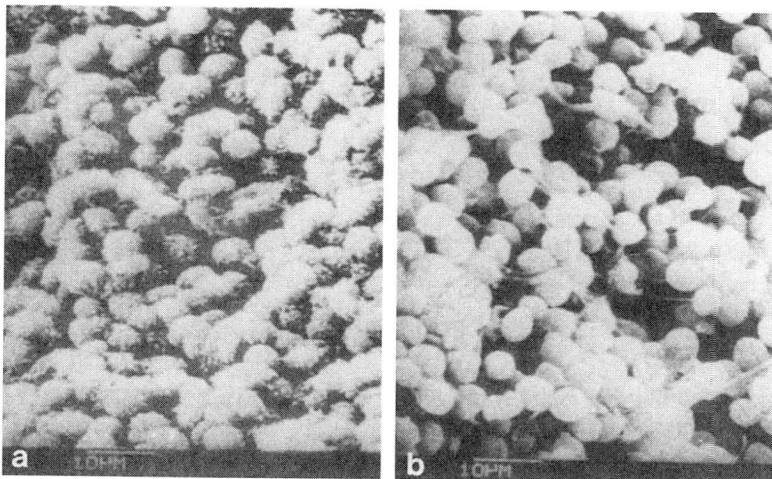

Fig. 4.3 a, b. Scanning electron micrographs of immobilized cells of *S. cerevisiae* [28]. **a** Cells entrapped in Ca alginate-gelatin composite beads. Evidence of budding. **b** Entrapped cells in matrix after leaching out of alginate and crosslinking with glutaraldehyde

bodies of methanol utilizing yeast (Kloeckera sp 2201). Formaldehyde and formate dehydrogenase were localized in the cytochromic region of the yeast [24]. Knowledge of the distribution of enzymes in the cell (intercellular or intracellular) is important in the planning of an immobilized cell fermentation, particularly in connection with the decision to use permeabilization and enzyme fixation techniques, especially when enzyme production is the objective of the fermentation.

Cell Viability. The most direct visual evidence of cell viability is an observed increase in cell population (see Fig. 4.2 [25]) [26, 27], a change in cell distribution [25], the appearance of buds, or evidence of budding (see Fig. 4.3 [28]) such as bud scars and birth marks on parent and daughter cells [25, 28] and cell division [12].

As observed [29] by SEM, *Pseudomonas putida* cells contained ill-defined boundaries after immobilization but soon changed into densely packed cells with spherical boundaries, and, when incubated with a suitable medium, showed signs of lysis after prolonged elution (21 days). Lack of viability may be detected as a decrease in cell density or the presence of cell lysis [12, 25, 26, 30], whose effects are evident in ruptured cells, or formation of slime, as well as by a change in cell distribution in the carrier (Figs. 4.4 [12], 4.5 [30], and 4.6 [31]). Observations on cell viability may also be made using phase-contrast microscopy [32].

Evidence of Immobilization. Direct evidence of cell immobilization is difficult to obtain by electron microscopy, because the sites of adherence between the cells and the carrier responsible for immobilization exist as a result of complex physical and chemical interactions which very often do not survive damage caused by the procedures of sample preparation and observation such as freeze-drying, sec-

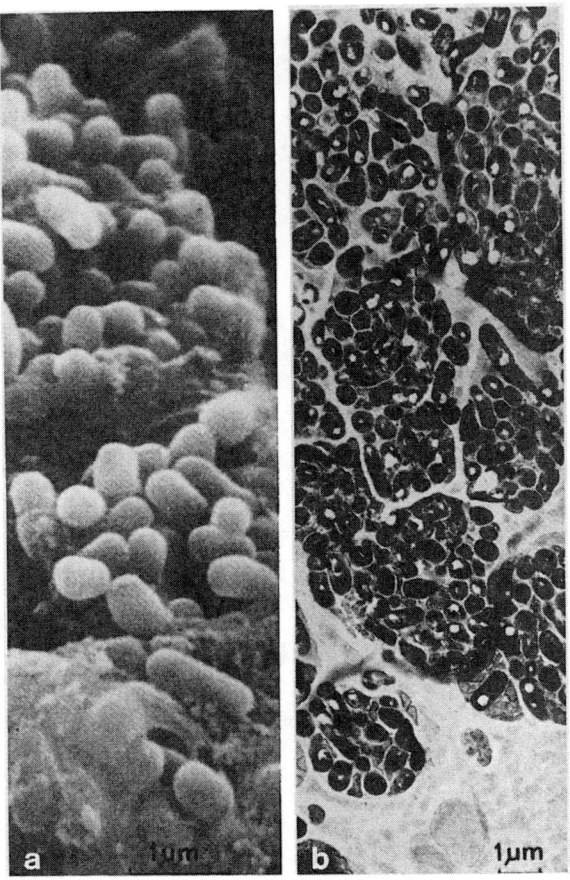

Fig. 4.4 a, b. State of cells on the surface and inside granules after 1–2 months of batchwise transformations [12]. **a** Ruptures in the gel film. **b** Ultrathin section of the granule subsurface layer

tioning, dehydrating, subjecting to high vacuum, defective metal coating and electron bombardment. Immobilizations by electrostatic attraction or by mechanical confinement are especially difficult since there may be no substance linking the cells to the carrier [33]. In one instance, the encapsulation of *Micrococcus denitrificans* by liquid membrane [32] has been observed to advantage by phase contrast micrography because the nature of the immobilization medium was not destroyed.

Evidence exists of cell attachment via cellular polysaccharide. For example, Brooker and Fuller have observed the immobilization of two strains of *Lactobacilli* on chicken crop epithelium [34] using cationic dyes of colloidal iron, ruthenium red or lanthanum nitrate instead of the routine heavy-metal staining (Fig. 4.7 [34]). *Lactobacillus* cells on crop epithelium fixed in Alcian blue showed no apparent cell-epithelium contact except for some loose material in the space. However, the use of the cationic dyes revealed filaments of surface carbohydrate

100

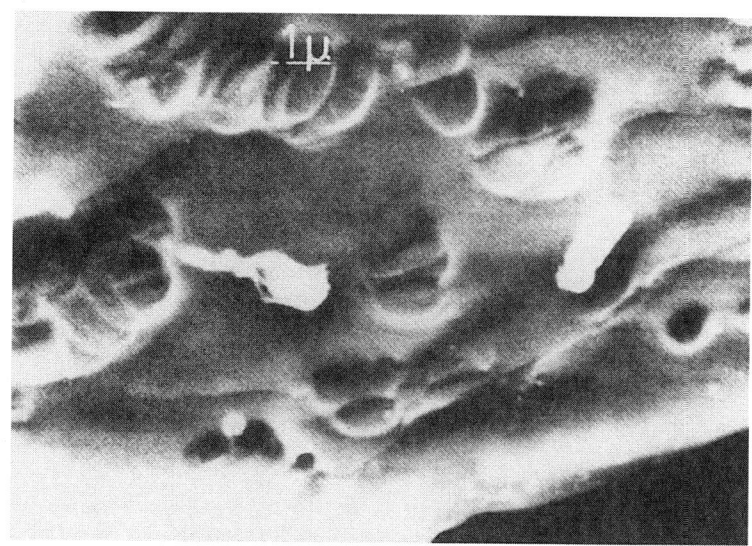

Fig. 4.5. Electron micrograph of immobilized *Alcaligenes faecalis* cells (7000 ×) [30]

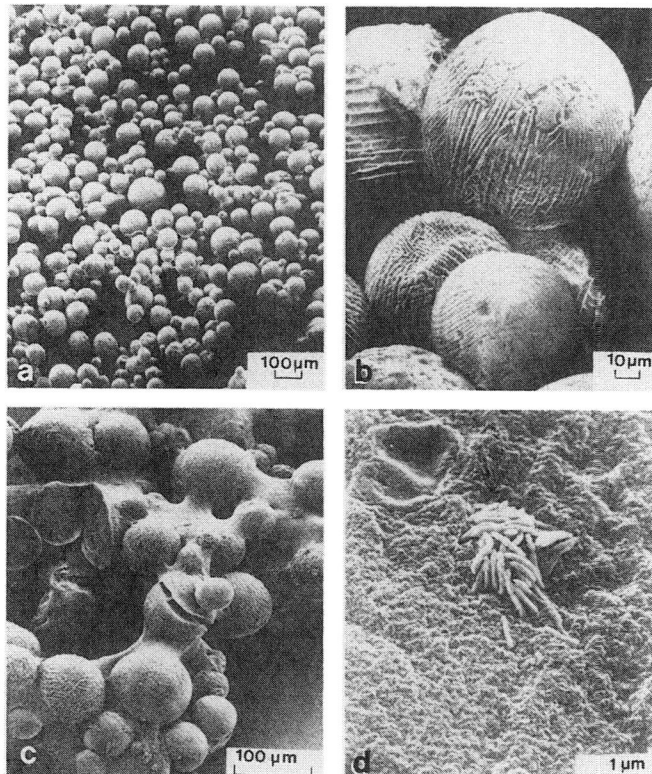

Fig. 4.6 a–d. Beads of Sepharose 4B [31]. **a, b** Before incubation with bacteria. **c, d** After 16 h cultivation of the Gram-negative bacterium

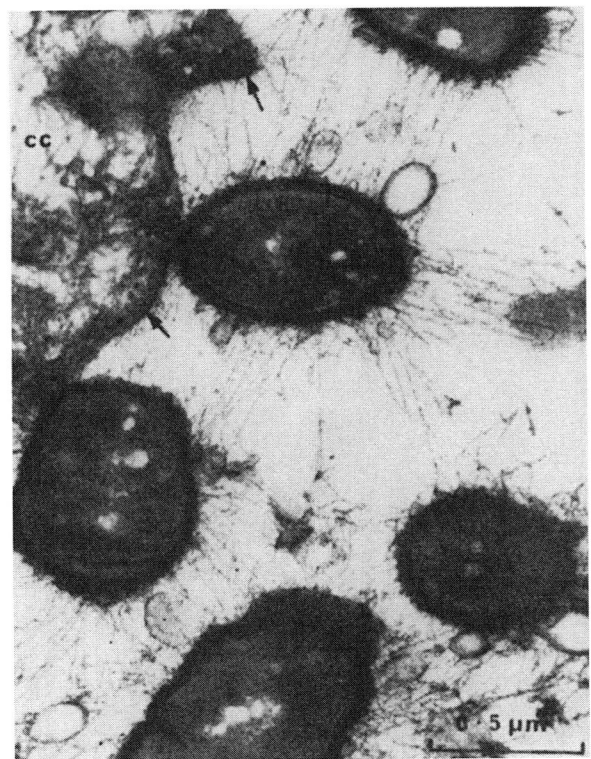

Fig. 4.7

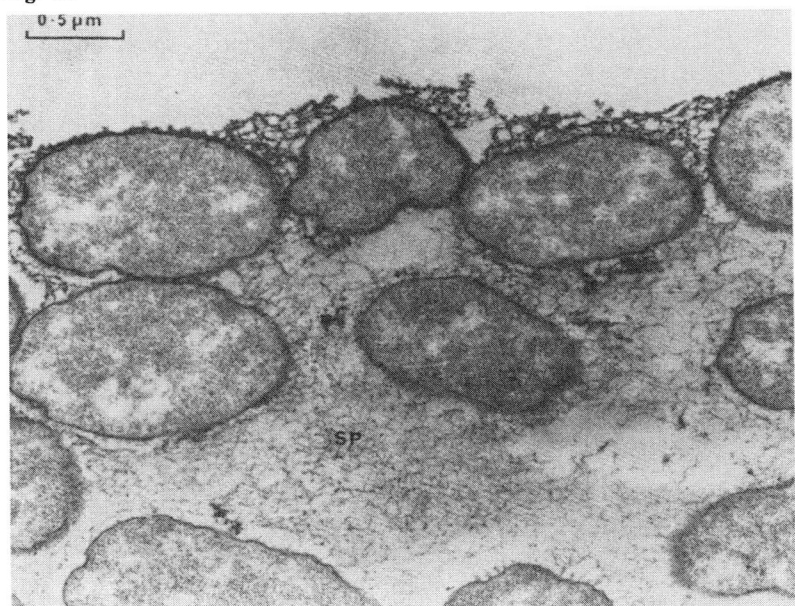

Fig. 4.8

102

material between the cell and the epithelium, as well as sites identified as arising from primary adhesion. The mode of adhesion to a MF-Millipore filter surface of cells of *Pseudomonas* sp. is depicted by electron micrographs of sections of these cells as consisting of a thin primary polysaccharide layer at the site of the cell-filter contact together with a more substantial secondary polysaccharide layer which replaces the primary layer to form a matrix for the cells in the colony (Fig. 4.8 [35]).

4.2 Properties of Enzymes in Immobilized Cells

4.2.1 Single and Multiple Enzymes

The chemical reactions performed by immobilized cells arise from either a single enzyme connected with the cells or from multiple enzymes acting in concert with the total metabolic process of the cells. In the first instance, so long as the enzyme is stabilized structurally and chemically in the cell, its activity will be conserved and the operational stability of the immobilized system will be high even though the cell may be non-viable or dead, as indeed is the case when a relatively harsh method of immobilization, such as those calling for covalent bond formation, is used. However, formation of products that involve the overall metabolism in the growth phase of the cell, including multiple enzymes and coenzymes, requires living cells. Examples of both categories of reactions have been compiled by MAXON and reviewed by CASIDA [36] and by MOSBACH [37].

4.2.2 Location of the Enzymes

Enzymes are either intracellular, extracellular or ectocellular, according to where they are usually located. An intracellular enzyme is located at sites inside a cell and may present diffusion problems in a fermentation reaction especially if high molecular weight substrates or products are involved. Very often the cell walls can be permeabilized by dissolution in some organic solvent or by a freeze-thaw cycle. Extracellular enzymes are enzymes produced by a cell and secreted into the environment through the cell wall, their chief function being to catalyse breakdown of large substrate molecules by hydrolysis reactions for absorption into the cell. Extracellular enzymes are also closely related to the formation of spores, the function of the cell wall and membranes and the genetics of the cell. An account of these relationships as well as the mechanisms and genetic control of the formation

---

Fig. 4.7. Lactobacillus attached to the crop epithelium treated with ruthenium red but otherwise unstained and showing bacterial filaments extending to other organisms as well as to the crop cell membrane. Arrows indicate areas of the crop cell membrane where surface staining material can be seen clearly (× 36 500) [34]

Fig. 4.8. Section from bacterial film showing the bacteria and their associated polysaccharides. Primary polysaccharide is eventually replaced by secondary polysaccharide which forms an intercellular matrix. The organism was grown in a peptone-yeast extract-aged seawater medium [35]

of extracellular enzymes is given by ERICKSON [38]. As an end product, extracellular enzymes are important industrially; however, as catalysts involved in immobilized cell fermentations, they may present problems of wash-out in a fermentor and may need crosslinking (by glutaraldehyde, for example) to become attached to the cell. Ectocellular enzymes are extracellular enzymes bound to parts of the cell membrane. Enzymes can be located in a cell by microscopy, using special staining techniques, or by chemical means. The operational stability of a fermentation reaction is based on the stability of enzyme activity after a given period of operation under known conditions of temperature, pH, etc. Operational stabilities of immobilized enzymes have been compiled by CHIBATA [39].

4.2.3 Operational Stability of Enzymes

Immobilized enzymes generally show higher operational stabilities than free enzymes, because after binding to a solid carrier optimal enzyme conformation and structure may be maintained. Immobilization may provide some protection for reactive sites on the enzyme against denaturization and provide easy access for co-enzymes and substrates. The operational stability of enzymes in immobilized cells often depends, in addition to the above conditions, on the state of the microbial cell as a living entity, and on the condition of attachment of the cell to the carrier. Perturbations of the physiological status of the cell, therefore, usually result in corresponding effects on the enzyme activity. Stability is commonly measured by the amount of activity after a certain period of operation and is most conveniently represented by the half-life, $t_{1/2}$, the time period after which half of the enzyme activity remains. Stabilities of enzymes are listed by CHIBATA [40] and CHIBATA and TOSA [41].

Some recent examples of enzyme stabilities in immobilized cells are listed in Table 4.2.

Kinetics
Fermentations using immobilized cells generally follow the Michaelis-Menten equation [20]:

$$V = \frac{V_{max}[s]}{K_m + [s]},$$

where

$$V = \text{initial rate},$$

$$V_{max} = \text{maximum rate},$$

$$[s] = \text{substrate concentration},$$

$$K_m = \text{the Michaelis constant}.$$

This equation is based on the steady-state attained in the concentration of a substrate-enzyme adduct SE in the formation of product p in the following reaction scheme

$$S + E \underset{k_{-1}}{\overset{k_1}{\rightleftharpoons}} E - S \underset{k_{-2}}{\overset{k_2}{\rightleftharpoons}} E + P.$$

During steady-state, the concentration of the substrate-enzyme complex SE is constant so that

$$\frac{d[E \cdot S]}{dt} = k_1[S][E] - (k_{-1} + k_2)[E - S] + k_{-2}[E][P] = 0$$

or

$$\frac{[E - S]}{[E]} = \frac{k_1[S] + k_{-2}[P]}{k_2 + k_{-1}}.$$

In the beginning of the reaction, very little product has developed, so that $[P] \cong 0$ and

$$\frac{[E - S]}{[E]} = \frac{[S]}{K_m},$$

where

$$K_m = \frac{k_2 + k_{-1}}{k_1}.$$

The total enzyme concentration $[E]_T$ is equal to $[E] + [E - S]$, so that

$$[E - S] = \frac{[S]([E_{tot}] - [E - S])}{K_m}.$$

Further, since the ratio $\frac{[E - S]}{[E_{tot}]}$ is equal to $\frac{V}{V_{max}}$, where V is the initial reaction rate and V_{max} is the maximum reaction rate,

$$[E_{tot}] \cdot \frac{V_0}{V_{max}} = \frac{[S]([E]_{tot} - [E - S]) \dfrac{V_0}{V_{max}}}{K_m}$$

or

$$V = \frac{V_{max}[S]}{K_m + S}.$$

Thus, initially in a reaction, when [S] is high,

$$[S] \gg K_m$$

Table 4.2. Operational stability in immobilized cells

Microbe	Carrier	Enzyme or enzyme action	Period of continuous fermentation	% Activity remaining	pH	T (°C)	Ref.
Azotobacter vinelandii	Anionic exchange cellulose	Nitrogenase	117 h	100%			[144]
Escherichia coli	Carrageenan gel	Gluthathione synthetase	8 days	50%		37°C	[117]
Escherichia coli	Glutaraldehyde	Immobilization of penicillin amidase	≧130 days		3.9	>45°C	[145]
Enterobacter aerogenes	κ-Carrageenan	2,3-Butanediol production	10 days	100%	7	30°C	[146]
Zymomonas mobilis	Ca alginate, κ-carrageenan	Ethanol production	800 h	70%			[147]
Bacillus sp.	Polyacrylamide gel	Production of bacitracin	~10 days	50%			[148]
Candida rugosa	Polyacrylamide gels	Fumarase	95 days	50%	8.5	30°C	[149]
Pseudomonas dacunhae	Carrageenan	L-Aspartase β-decarboxylase	46 days	58%		37%	[150]
Clostridium acetobutylicum	Ca alginate gel	Butanol production	1000 h				[151]
Zymomonas mobilis	Ca alginate	Ethanol production	384 h				[152]
Brevibacterium flavum	Collagen	Glutamic acid production	5–10 days	100%			[153]
Streptomyces fradiae	Polyacrylamide gel	Protease production	30 days	50%			[154]
Arthrobacter globiformis	Polyacrylamide	3-Ketosteroid-Δ'-dehydrogenase	(1)	72–92%			[12]
Actinoplanes missouriensis	Cellulose beads	Glucose isomerization	60 days			60°C	[155]
Kluyveromyces fragilis	Calcium alginate gel beads sodium alginate		At least 1 month				[156]
Daucus carota	Ca alginate	Bioconversion of gitoxigenin to 5β-hydroxygitoxigenin	>30 days				[157]
Rhodospirillum rubrum and *Klebsiella pneumoniae*	Agar	Production of hydrogen from organic substrates	1000 h	50%			[158]

Organism	Carrier	Process	Time	Yield	pH	Temperature	Reference
Saccharomyces cerevisiae		Production of ethanol	>2 months				[159]
Nitrosomonas europaea	Polyelectrolyte complex	Ammonia-oxidizing	2000 h				[160]
Corynebacterium glutamicum	κ-Carrageenan matrix	Glutamate production	30 days				[161]
Pleurotus ostreatus	Chitosan	Production of 6-aminopenicillanic acid	25 days	50%			[162]
Erwinia rhapontici	Alginate gel pellets	Formation of iso-maltose from sucrose	8600 h	50%			[22]
Lactobacillus delbrueckii	Calcium alginate gel beads	Production of L-lactic acid from glucose	100 days	50%			[163]
Enterobacter aerogenes	Photo-crosslinkable resin prepolymer	Production of adenine arabinoside	35 days	100%			[138]
Streptomyces sp. and *S. tendae*	Calcium alginate gel	Production of tylosin and nikkomycin	350 h				[164]
Leuconostoc oenos ML 34	Calcium alginate gels	Malate decarboxylase (I)	36 days	60%			[165]
Saccharomyces cerevisiae	Ca alginate	EtOH production	7–8 days	100%			[18]
Providencia sp.	Alginate	L-Amino acid oxidase	>1 month with reinoculation				[166]
Kluyveromyces marxianus	Alginate beads	Production of ethanol	12 days	92%			[167]
Clostridium sp.	Photo-crosslinkable prepolymers	Hydrogenation of Δ^2-enoates or aldehydes	10 days	50%			[168]
Lactobacillus delbrueckii	Ca alginate	Production of L-lactic acid	~100 days	50%			[169]
Trichoderma sp.	Calcium alginate beads	β-Glucosidase	1000 h	50%			[170]
Penicillium urticae	κ-Carrageenan	Patulin production				50°C	[171]
Schizosaccharomyces pombe	Flocs	Production of EtOH	2 months				[172]
Escherichia coli	Starch	β-Galactosidase	8 days	100%	5.5–8.0		[173]
Escherichia coli	Polyacrylamide beads	Tryptophan synthetase	50 days	80%		8–55°C	[174]
Corynebacterium sp.	Photo-crosslinked gels	Steroid 9α-hydroxylation system	5 days[a]	100%			[133]

[a] Batchwise process.

and

$$V_0 = \frac{V_m}{K_m}[S]$$

and the reaction follows first-order kinetics. When $V_{max} = 2V_o$, K_m has the numerical value of [S]. As substrate is depleted,

$$[S] < < K_m$$

and

$$V_o \cong V_{max} = \text{const},$$

that is, the reaction becomes zero order. In general, the reaction is of the mixed first order and zero order. The Michaelis constant is an index of the reactivity of the enzyme towards the substrate in an enzyme catalyzed reaction and its relationship to substrate concentration determines the order of the reaction.

The Michaelis-Menten equation suggests that the enzymes compete to occupy a limited number of active sites in the substrate to form a complex. As soon as all the sites are filled, the concentration of this complex is constant and the reaction proceeds at the constant maximum rate. Inhibitors are substances that either compete and occupy reactive sites with the enzyme (competitive inhibition) or in addition to occupying a site, further deactivate other active sites towards enzymes (non-competitive inhibition). A compilation of Michaelis constants for immobilized enzymes is provided by CHIBATA [42].

The kinetics of immobilized cells can be treated as a perturbation of the kinetics of free enzymes in terms of such effects as enclosure in a cell, attachment of that cell to a solid carrier and diffusion limitations imposed by the surroundings. If the fermentation is carried out in a reactor, then the characteristics of the reactor will further modify the kinetics. There have been many reviews of the above effects on immobilized cell kinetics [43–45]. The following is a list of modification effects on the kinetics of free enzymes:

(a) Modification by the cell environment:
   pH
   temperature
   substrate concentration
   transfer of ions, substrates [46]
   inhibition by substrates and products [46]
   mass transfer effects [46]
   diffusion across cell wall and cytoplasmic membrane [44]
   cell division [44]
(b) Modification by the act of immobilization [44]:
   different conformation requirement and steric effect after immobilization
   concentration variations of substrates, products in the carrier diffusional effects, both across boundary film [47, 48] and into the pores [49, 50] of the carrier.

(c) Modification by the carrier:
microenvironments in the immobilization system [51].

4.3 Morphology of the Carrier

The properties of solid carriers for immobilized enzymes have been extensively reviewed [52, 53]. Although these reviews are primarily concerned with enzyme immobilization, they are also applicable to microbial cell immobilization with certain qualifications, the most important of which are that some fermentations are carried out by viable or growing microorganisms or by multiple enzymes, and

Table 4.3. Electron microscope studies of carriers used in cell immobilization

Carrier	Mode of study[a]	Pore size	Fine structure	Ref.
Polyacrylamide gel	SEM		Effects of crosslinking, monomer concentration	[59]
Agarose beads	SEM		Furrows and depressions, slime due to degradation	[31]
Sepharose 6B (ferritin stained)	SEM	42 nm	Shell-like layer of covalently bound ferritin in Sepharose	[55]
Agarose beads	SEM	0.3 μm	Filaments of diameter 20 Å in bundles, pores > 0.3 μm, sponge like structure	[57]
Cellulose beads	SEM	1 000 Å		[56]
Bovine serum	SEM		Sectional surface: homogeneous structure	[60]
Albumin foam	TEM		Fractured surface: humps and holes	
κ-Carrageenan	SEM		Surface microinterstices during stationary growth of yeast cells	[25]
4-Isothiocyanato-styrene/acrylic acid/1,4-divinylbenzene	SEM		Fissure due to osmotic effects	[175]
Controlled pore glass	SEM	3 000 Å	Interconnected pores	[74]
Photo-crosslinkable resins	SEM			[176]
Cellulose triacetate fibre	SEM		Porous, microfibrillar structure	[54]
Polyacrylamide gel	SEM		Rupture of gel film 2 months after incubation	[12]
Oxirane acrylic beads 2878C (Röhm GmbH)	SEM	0.1–2.5 μm (1 000–2 000 Å)		[177]
Albumin, polyurethane, gelatin, calcium alginate, photo-crosslinkable resin, carrageenan	SEM			[178]

[a] SEM: scanning electron microscopy.
TEM: transmission electron microscopy.

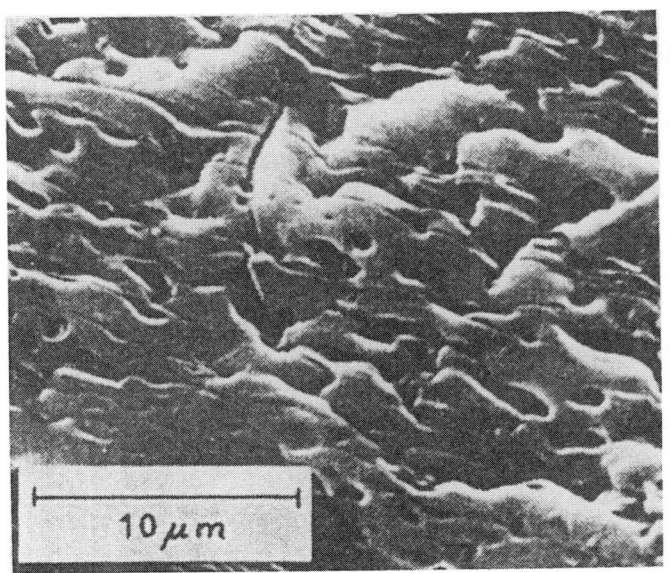

Fig. 4.9. Scanning electron micrograph of cellulose triacetate fibre [54]

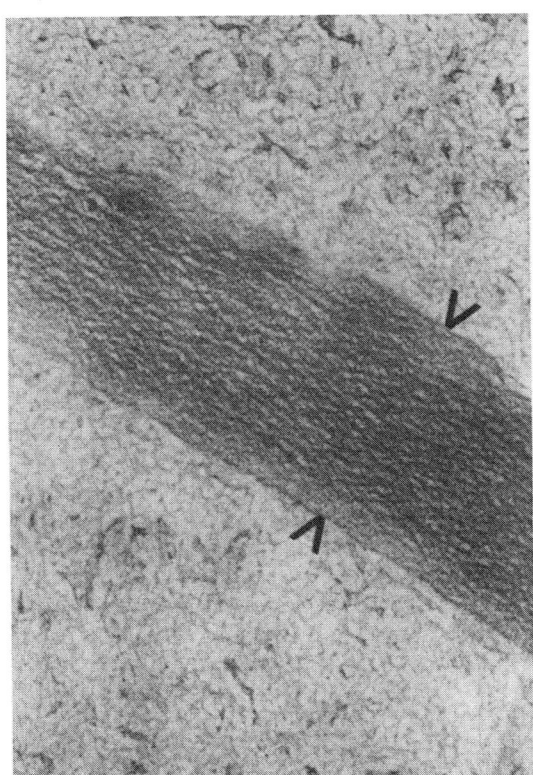

Fig. 4.10. Non-activated Sepharose treated with ferritin (*blank*). Final magnification: 29 440-fold. Gelatine embedding. The positively stained gelatine-filled space between two gel beads is indicated by arrowheads [55]

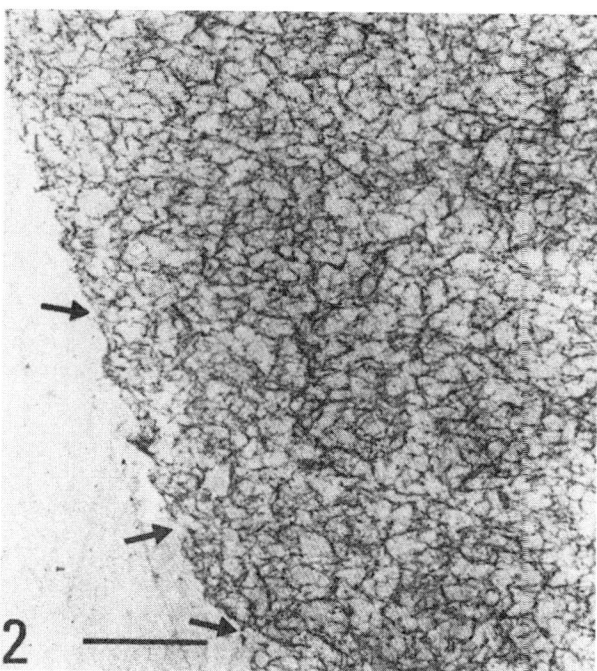

Fig. 4.11. Electron micrograph showing the outer edge of a Sepharose 4B bead embedded in Epon. Arrows indicate openings of the pores or channels into the outer surface. Calibration, 1 μm [57]

that whereas enzymes are of molecular dimensions, microbial cells are of the order 1000 Å. This chapter reviews recent work (since the late 1970's) on the subject of properties of the carrier system in connection with immobilized cells.

4.3.1 SEM Observations

Carriers for immobilized cells have been studied by electron microscopy (Table 4.3) with the objective of determining morphological characteristics such as pore shape [54], pore structure [55], pore size and distribution, and fine structures. The pore size distribution of carriers is normally determined by the mercury porosimeter but can also be determined by SEM [56] and TEM [55].

Shrinkage of cellulose beads on drying, a common problem in sample preparation, can be avoided by using critical point drying. Because of the great depth of field of SEM, objects can be viewed as a whole, so that SEM is often used in monitoring the physical condition of the carrier (see Figs. 4.9 [54], 4.10 [55], and 4.11 [57]), especially in detecting damage to the carrier due to abrasion, shear forces, osmotic pressure [58], microbial degradation [31], defects in polymerization [59] and defects in crosslinking [59] (see Fig. 4.12). The interior of solid carriers can be observed by SEM on the section surface or fracture surface of the

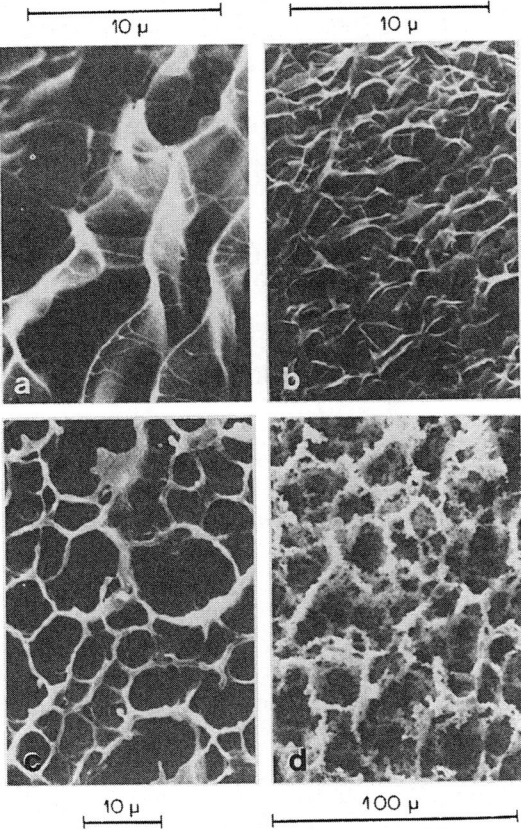

Fig. 4.12 a–d. Scanning electron microscope images of polyacrylamide gels of constant monomer concentrations (10% w/v) and different stages of cross-linking. **a** Image of a gel without cross-linker that reveals parallel leaflet structures at low magnification levels. These leaflets, as shown here, are connected by slender fibrils. **b** Image of a 0.2% cross-linked gel, which does not reveal the straight parallel organization of the pure monomer gel but shows parallel organization at higher magnification levels. This image can be suppressed in favour of a more cellular structure, as shown here. **c** Image of a 10% cross-linked gel that became fully opaque during polymerization. The image reveals a decreasing organization of the cellular structures. Membrane defects in the form of round holes may be related to decreasing sieving capabilities of highly cross-linked gels, and bulblike structures are likely to be clusters of cross-linking "knots". **d** Image of pure polymerized cross-linker (initial concentration 2% w/v) at lower magnification. The "gel" reveals the image of cellular organization, but at higher magnifications there is only a random aggregation of spherical units [59]

solid (Fig. 4.13 [60]). For polyacrylamide gel, no significant structural difference was detected [59] on these two surfaces.

4.3.2 Particle Morphology

The size, size distribution and shape of particulate solid carriers (distinguished from membrane and fibre carriers) play important roles in determining the bulk

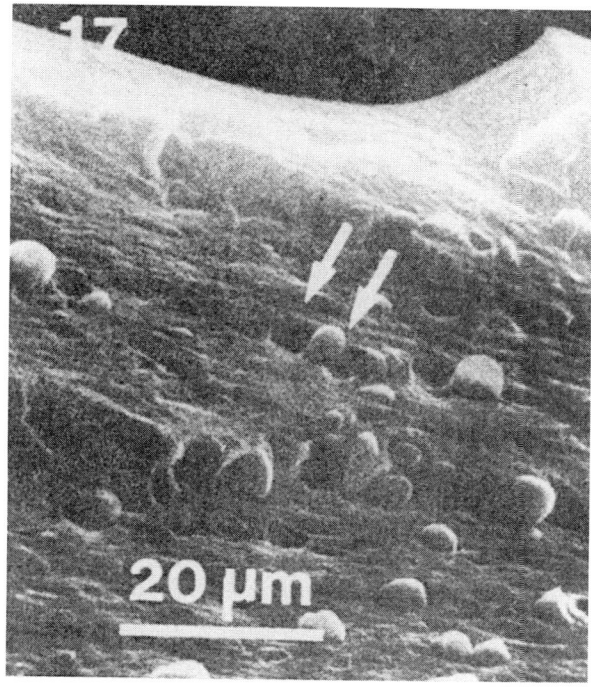

Fig. 4.13. Hand-fracture of the internal part of the foam polymer visualized by scanning electron microscopy. The presence of humps and holes should be noted [60]

characteristics of immobilization systems, especially the kinetics of fermentation and the design of reactors. Solid carriers, after swelling in water, are usually crushed or mashed through screens, if not already in granular form, and sieved to obtain a fraction of a known range particle size. Sieving is carried out on screens submerged in water or in an organic solvent that is chemically inert to the components of the carrier.

Particle size analysis can be performed by standard procedures, giving particle size distributions and also the particle dimensions of the sample. Size analysis is usually carried out by microscopy or the Coulter counter [61]. The size characteristics of the carrier influence the course of the immobilized fermentation [62–64]. The particle size of a carrier determines the surface area of that carrier, the amount of loading [65] and ultimately the activity and the rate of the fermentation. The size of the carrier particles determines the level of agitation necessary to keep them suspended in the liquid, and also the pressure drop in a fixed bed or a fluidized bed reactor.

Particle shapes affect such behaviour as hydrodynamic interaction with fluids, degree of abrasion, behaviour in packing, sedimentation, filtration and fluidization. Attempts to define quantitatively these relationships include various methods of expressing irregular shapes of the particles in terms of parameters of solids of well defined shapes [66]. Whenever possible, carriers are used in the spherical bead form, which converts the kinetics of the fermentation to the kinetics of heterogeneous catalysis.

A certain flexibility in particle shape and size may be obtained by designing carrier particles from wire mesh [67]. These wire mesh particles, in the form of spheres from a single strand of stainless steel wire, are found to be especially suitable for accumulating living biomass of several types of cells in a variety of reactor types. Polyhedral shaped carriers used to immobilize *Aspergillus* fungi in the production of enzymes have been found to float freely in the fermentation mixture for enzyme production [68].

4.3.3 Pore Size

The pore size of solid carriers for immobilization has a wide range, from non-porous substances such as glass to the very porous organic structure of cellulose in wood.

Pore size and distribution in a carrier can be determined by several methods [69, 70]. In general, pores in a carrier are divided into those open at both ends and those open at one end only. In microbial immobilization, both kinds of pore are of interest and pore size is commonly determined by the mercury porosimeter. In this method, the solid with pore radius r is placed under vacuum together with mercury. Pressure is then increased incrementally until a threshold value is reached at which mercury is forced into the pore to a certain depth. When mercury of viscosity, μ, and contact angle, θ, with the carrier is forced under pressure, p, into a pore, of radius, r, the following relationship holds

$$p \cdot r = 2\mu \cos\theta.$$

In general mercury will not penetrate into the capillary pore to any great extent until the pressure is increased to a threshold value and thereafter penetrates to a smaller extent beyond that pressure. From this expression, a distribution function can be obtained in terms of the parameters, p, r and the volume of pores containing mercury. A plot of p vs. this volume gives a curve from which the pore distribution can be obtained [52].

Two methods of permeability determination, based on ammonium chloride and on the reactivity of encapsulated urease, have been applied to nylon membranes [71]. The pore size distribution in a carrier can usually be designed within certain limits. The most effective and precise method is that of preparing glass particles with controlled pore size distributions [72–74] in the 10^2 to 10^3 Å range. The method is based on the phase behaviour of two liquid components with respect to temperature. In this system there is a temperature above which the two liquid components are miscible, but below which mixtures not predominantly rich in either component will give two immiscible phases consisting of mixtures of the two liquids in different proportions. In practice, the three components B_2O_3–SiO_2–Na_2O are used. In the molten state, their high viscosity effectively disperses the emergent phases into microscopic regions of different sizes and shapes depending on the stage of the cooling at which they are formed. Dissolu-

114

tion of B_2O_3 and Na_2O by dilute hydrofluoric acid or dilute sodium hydroxide produces a lattice-like structure consisting of the original silica. The possibilities of preparing glass carriers with a fairly narrow range of pore size were explored in a study of the optimal pore size for biomass accumulation [75].

That the optimal pore size is related to the mode of reproduction of the micro-organism involved can be explained by a model containing pores in which a layer of microorganisms is attached to the wall and a space in the center is available through which a microorganism can pass. Based on this model, the optimal pore sizes for a microorganism of the major dimension of, say, 1 μ and reproducing by fission and budding would be 5 μ and 4 μ, respectively. The size for reproduction by spores is considerably higher, up to 16 times the size of the spores [76]. Experimental measurements provide fair agreement with the model. The discrepancies are due to the non-uniformity of microorganism sizes, surface effects on carriers, and the particular phase in the growth cycle of the cells.

The physiology of *Saccharomyces cerevisiae* cells covalently crosslinked on controlled pore glass, as measured by the rate of O_2 uptake, CO_2 evolution and cell size, has been found to be similar to that of free cells, with the exception of a shorter generation time [9].

Control of carrier particle size can be effected by several methods, one of which involves dispersing a hydrophobic monomer in cold water followed by polymer-ization. A modification of this procedure is to use a dispersed hydrophilic monomer phase formed by salting out. Mixtures of the monomers 2-hydroxy-ethyl methacrylate and 2-hydroxyethyl acrylate, dry cells and a salt solution were shaken to salt out and disperse a monomer phase. Cooling to -78 °C and expos-ing to γ-radiation gave fine particles containing immobilized cells. The particle size has been shown to be determined by the monomer concentration, salt concen-tration and irradiation temperature [77]. A similar procedure was used by ZIOMEK et al. [65] to prepare polyacrylamide particles for the immobilization of *Desul-phovibrio desulphuricans*. The specific activity of these immobilized cells has been found [65] to approach that of free cells as the particle size of the carrier decreased (at 0.05–0.1 mm particle diameter, the specific activity was 92.0% of the free cells).

A mixture of cordierite and alumina with known pore size has been used to sta-bilize a number of bacterial cells for prolonged storage after freeze-drying [78].

4.3.4 Porosity

Biological materials with a wide range of porosity can be used as carriers for im-mobilization, for example, yeast cells on sugar-cane bagasse pith [79], and on a composite carrier of tuff granules and gelatin [80] and microbial mycelial mats [81] and pellets [82]. The porosity of calcium alginate gels can be increased by treating with agents such as phosphates which form complexes with calcium (see Chap. 2) or increased by replacing calcium with other divalent cations such as barium or strontium [83]. Porous material in the form of hydroxyapatite has been mixed into the carrier κ-carrageenan. The resulting composite gel has been found to provide higher loading, increased viability, and increased ethanol yield for yeast cells [84]. Control of porosity is possible in using as carriers wire mesh pellets formed by crushing a single strand of stainless steel wire [67].

4.3.5 Compressibility

Solid carriers containing immobilized microorganisms are used in several kinds of reactions in fermentation, in forms such as packed bed, fluidized bed, hollow fibre, membrane, etc., operating in the batch-wise or continuous mode. The use of carriers that are compressible or deformable introduces complications in the treatment of these fermentation reactions, mainly in the occurrence of creep which can induce the porosity of the column.

Treatment of the hydrodynamics of packed material usually assumes as a zeroth approximation spherical uniform incompressible solid particles and laminar non-viscous flow. In this case, the pressure drop across a section of the column is defined by the Hagen-Poiseuille equation and is constant for fixed values of column height, linear flow and viscosity of the liquid in the reactor [85, 86].

Behaviour of carrier particles that are compressible in a packed column is described by BUCHHOLZ and GÖDELMANN [64]. In general, the corresponding pressure drop across a section of the column is no longer constant but is shown to vary nonlinearly depending on, among other things, particle size, column height, flow rate, and the history of the packed column.

A theoretical treatment of the pressure drop in a compressible column of particles has been developed by BUCHHOLZ and GÖDELMANN [87], using as analogy methods developed for filtration through compressible layers [88]. Based on the assumption that the elastic modulus of the spherical particles is defined as the fraction of radius compressed per unit of pressure, the pressure drop at each succeeding layer of spheres was calculated from the Hagen-Poiseuille equation. When applied to the carrier Sepharose CL 6B, at a pressure of 200 mbar, the pressure drop at different column heights was predicted up to the point at which compression is sufficiently large that all pores are blocked, as indicated by an asymptotic rise in the curve.

Methods for determining the flow resistance of carrier particles are described by KLEIN and KLUGE [89]. The compressibility of alginate spheres can be changed by varying the concentration of alginate and the microorganisms in the carrier and the temperature. The effects of creep under compression can be reduced by maintaining a high flow rate up the column. A problem associated with some porous, compressible carriers, for example, agar or κ-carrageenan, is bursting caused by gas evolved in the fermentation. Bursting can be prevented by incorporating polyacrylamide into the lattice of agar or κ-carrageenan carriers to provide higher mechanical strength [11]. Carrier beads are first treated with the monomer which enters the beads by diffusion. The monomer is then polymerized in the bead on treatment with an initiator. Another composite gel is made by incorporating polyethyleneimine into alginate [90]. Both of these gel composites are reported to be superior carriers with a high mechanical strength, thermal stability and a stabilizing effect on the immobilized microorganisms. Polymer particles with a marshmallow-like composition are formed by irradiating the capsules formed by spray-

116

ing a mixture of the enzyme cellulase and polystyrene in glycidylmethacrylate into ethanol at $-78\ °C$, followed by soaking in acetone to swell the capsules. The mechanical strength of alginate pellets can be increased simply by using higher concentrations of alginate and calcium [91].

The addition of various mineral solids such as zeolites, silica gel, and glass to polyacrylamide gel may produce a reinforced gel carrier [92] having improved hydrodynamic and mechanical properties.

4.3.6 Abrasion and Rupture Resistance

During a fermentation, the carrier particle is usually subject to forces which tend to fracture it, for example, shearing forces arising from stirring, fluidizing or from collision among the particles and reactor wall. In addition, gases produced by microorganisms immobilized in a carrier particle may expand, thereby rupturing the particle. Finally, the carriers may be degraded by the osmotic action of the medium. Most inorganic and plastic carriers suffer little damage due to abrasion. However, marine algal carriers including agar, alginate and κ-carrageenan are susceptible to considerable wear due to abrasion and rupture from gas formation.

The parameters defining the shear stress of an impeller in the mixing process in a fermentor are discussed and investigated experimentally by BUCHHOLZ [93]. The most important contribution to the shear stress is the angular speed of the impeller blade tip. With carriers containing immobilized cells, abrasion as measured by nephelometry was shown to depend on the cell density in the carrier [94]. A similar effect has been observed between the shear due to shaking and the rate of cell leakage, which, however, could be reduced by an increase in the potassium ion concentration in the medium, at the expense of cell viability [95, 96].

Fermentation using filamentous microbial growth is particularly susceptible to damage by shear forces [82]. One technique that has been shown to mitigate this effect is the use of wire mesh particles as carriers in the growth of biomass [67]. These carriers have been demonstrated to be suitable for batchwise or continuous fluidized bed reactors.

Gas production by immobilized microorganisms has been formulated as a zero order reaction, with diffusion of gas as the rate determining step. Using this kinetic model, KROUWEL and KOSSEN [97] derived conditions under which no gas bubbles are formed inside carriers of different geometry.

In general, reinforcement of carrageenan, and alginate gels – for example, reinforcing agar or alginate gels with polyacrylamide [11] – is effective in preventing gas rupture.

4.3.7 Thermal Properties

Carriers used in fermentation with immobilized cells are usually affected by temperature to a lesser degree than the cells or enzymes themselves. Among these carriers are proteins such as collagen and gelatin, and polysaccharides such as calcium alginate, agar, agarose, pectin and carrageenan. Many of these solids readily dissolve in boiling water and partially dissolve at lower temperatures. Their gel structure can also be damaged. In addition, high temperatures may intensify the consequences of gas eruption of the gel. Expansion of the carrier following a temperature increase may deform the carrier to such an extent that attachment of the microorganisms or the enzyme is weakened. As the temperature is increased, the equilibrium position of the adsorption process will generally be shifted to favour the desorbed species.

4.3.8 Surface Properties

The mechanisms of cell adhesion have been subject to intense experimental and theoretical investigation. Recent reviews on this topic [98, 99] concur that, at present, only a semi-quantitative understanding of these processes is possible pending more refinement of theory and accumulation of more relevant experimental data. Since the behaviour of free-flowing microorganisms adhering to another microorganisms or to a carrier surface has an immediate analog in the coagulation of dispersed particles in a colloidal system [100], microbial adhesion is often studied using theories developed in colloid chemistry, with appropriate modification.

The energy due to electrostatic interaction in this system as the two surfaces are brought closer to each other is formulated on a statistical basis using the Boltzman distribution. Under the assumption of constant potential or constant charge on the surfaces, as well as the geometry of a spherical particle and an infinite planar surface, it is possible to calculate the energy of interaction of the system as the distance of the two surfaces is gradually decreased. Under physiological conditions, LIPS and JESSUP [100] have found that as the inter-surface distance decreased, the electrostatic interaction becomes increasingly more attractive or repulsive depending on whether constant surface potential or constant surface charge is assumed.

In general, two broad categories of interactions are involved when two surfaces approach in an aqueous medium. The first is the electrostatic interaction between the counter ion layers near each surface. This counter-ion layer arises as a result of the electrostatic attraction of ions of the opposite charge by the net charge on the surface. The electrostatic interactions are mostly conservative and repulsive and operate at long ranges of several nm. If other dispersive interactions such as van der Waals forces and attraction due to hydrogen bonding are taken into account, the interaction of the system would be represented by a moderate potential minimum at distances greater than several nm, a stiff potential barrier at distances of close approach (about 1 nm) and stronger attractive forces at a deep potential minimum at distances less than 1 nm. The mechanism of microbial adhesion can be explained on the basis of an initial weak and reversible attachment several nm away, mainly due to electrostatic attraction, hydrogen bonding and ionic bonding. This attraction is countered by Brownian motion and steric hindrance. Closer approach of the entire microorganisms towards the surface of the carrier is not possible due to the high potential barrier. However, probes of polysaccharides on the surface of the microorganism are able to penetrate this barrier and establish

118

firmer attachments with the carrier surface at distances corresponding to the deeper potential minimum, estimated as being 100 kT or 4×10^{-12} ergs for bacterial cells [101]. At short ranges of separation, it becomes increasingly difficult to determine the potential of the immobilization system.

The situation at short ranges during adhesion can nevertheless be studied empirically from the free energy of the surfaces involved in the cell immobilization. This free energy is experimentally related to the angle of contact [98]. The result of this approach is the division into two sets of carriers of high and low energy. Carriers of high energy form stable but selective attachments with microorganisms, and carriers with low energy form weak but more non-selective attachments. Another empirical approach concerns the electrostatics of the adhesive surface near the shallow potential minimum. This minimum is related to the net surface potential or zeta (ζ) potential, a quantity used in electrophoresis to measure the mobility of the charged species in an electric field. The experimental measurement of the zeta-potential of yeast cells as a criterion for cell attachment is detailed by THONART et al. [102] and reviewed by GERSON and ZAJIC [98], DANIELS [103], MARSHALL [104], ATKINSON and DAOULD [105], and BORIES and RAYNAL [106]. The adhesion of *Saccharomyces cerevisiae* cells to sawdust, both having a zeta potential of -45 mV, has been found to be improved by addition of cationic gelatin which changed the zeta-potentials to -15 mV and -35 mV, respectively [107].

4.3.9 Chemical Properties

The chemical properties of the carriers can be discussed in terms of their inherent chemical nature and their reaction to the immobilized cells. In general, carriers for immobilization fall into one or more of the following groups: inorganic, natural polymers, synthetic polymers and crosslinking agents (Table 4.4). These carriers are sometimes combined to produce composite carriers for immobilization; the resultant chemical properties are to some extent a combination of those from the constituents. Certain chemical reactivities of carriers arise from the presence of impurities, especially those that have toxic or inhibiting effects on microorganisms or enzymes, for example, metal ions in inorganic carriers or in some gel carriers, and polymerization initiators, monomers and crosslinking agents in synthetic polymers. Although almost the same carriers are employed in both enzyme immobilization and cell immobilization, a much wider range of chemical modification is possible with the former. Thus for each type of carrier there are many chemical reagents that can be used to derivatize the carrier substance in order to introduce other reactive functional groups that will form covalent bonds with enzymes. Compared to cells, enzymes have better defined and a wider spectrum of chemical reactivity. Bonding is not always possible with microbial cells because many of the chemical reagents involved are cytotoxic. The chemistry of derivatizing various kinds of solid carriers as well as their modes of bonding to enzymes has been extensively reviewed [43, 108].

Some of the enzyme methods have been extended to cell immobilization and may be discussed in light of recent developments. In general, the enzyme activity,

Table 4.4. Classes of carriers used in cell immobilization

Inorganic, mineral	Silica (and hydroxy forms) – glass ceramics, soil, sand	
	Alumina	
	Titanium	
	Zirconia (and hydroxy forms)	
	Bentonite	
	Cordierite	
	Diatomaceous earth	
	Hydroxyapatite	
Polymers, natural	Proteins	albumin
		collagen
		gelatin
	Polysaccharides	agar, agarose
		alginate
		pectate
		κ-carrageenan
		cellulose (natural fibres
		and derivatives, ion exchangers)
Polymers, synthetic	Polyacrylamide	
	Photo-crosslinking polymers	
	Polyurethane	
	Polyethylene	
	Polypropylene	
	Polyester	
	Polystyrene	
	Polyphenylene oxide	
	Epoxy resin (polyvinyl chloride)	
	Polyvinyl alcohol	
	Silicone	
	Ion exchangers	
Crosslinking agents	Glutaraldehyde	
	Tannin	

half-life, product yield and sometimes the rate of reproduction of immobilized cells are increased compared to free cells. As previously noted, various theories have been proposed to account for this result.

In summary, the chemical influences of most of the solid carriers are generally beneficial to the enzymes or microbial cells in a fermentation; nevertheless, the presence of impurities in the form of monomer preservatives, serves, copolymers, etc. may result in effects on the fermentation which are difficult to assess. Very often reagents in the form of electron acceptors, permeabilizers, stabilizers, and antibiotics are added to the fermentation mixture. The effects of these must be considered.

Inorganic Carriers. As a class, mineral or inorganic carriers are fairly inert chemically, under normal conditions. However, carriers of glass, especially when finely divided, will react rapidly with even dilute base solutions and dissolve. Silicates have been silanized by reacting with amino-alkyl alkoxysilanes, which introduce amino-functional groups to the solid, which is then further derivatized to introduce reactive groups that can form covalent bonds to enzymes [108]. In microbial cell immobilization, the most often used reagent for crosslinking is gluteraldehyde, a dialdehyde one end of which condenses with the amino groups introduced into the silicate solids, leaving the remaining aldehyde group free on the other end to bind covalently to microorganisms. An example of this binding action is the use of silica beads or brick particles derivatized with γ-aminopropyltrimethoxysilane and glutaraldehyde to covalently bind cells of *Saccharomyces uvarum,* thereby increasing the generation time by a factor of 2–4 [3]. A similar effect was reported in the immobilization of *S. cerevisiae* on controlled pore glass [9].

Other procedures of crosslinking involve the use of alkylamine glass coated with zirconia and then treated with gluteraldehyde [109] and the binding of adsorbed cells on glass by glutaraldehyde [110].

Protein Carriers. Generally, collagen, gelatin and albumin are soluble or dispersable in water under different conditions. Albumin is the most easily soluble to give a gel and can be salted out from a concentrated solution of ammonium sulphate.

In the dry solid form, both gelatin and collagen absorb water. Collagen is denatured in boiling water to produce gelatin, which when dispersed in water exists as a gel or a solid depending upon temperature. Both gelatin and collagen are soluble in dilute acid solutions such as acetic acid. The mechanical strength of protein carriers is increased by crosslinking with glutaraldehyde [111, 112].

Polysaccharide Carriers. Many polysaccharides derived from plants are used for immobilization. From various species of sea-weeds and algae, the carriers carrageenan, agar and alginic acid are extracted. Pectin occurs in the rind of citrus fruits, and chitin in fungi, yeasts and the exoskelton of marine anthropods. Starch occurs in the cell wall of plants and starch in seeds as well as in roots of plants. Agar is a complex polysaccharide containing alternating α-(1→3) and β(1→4) linkages from which the neutral fraction agarose and a sulphate-containing fraction, agaropectin, are isolated. A 1.5% aqueous solution of agarose gels in the temperature of 42 °C to 15 °C. The purity of agarose is indicated by the percentage of sulphate remaining. Because the sulphate containing agaropectin does not gel, the gel-point of agar (1% aqueous solution) is slightly lower than that of agarose.

agarose

Alginic acid is a straight chain polymeric acid with anhydro-D-monouronic acid residues linked in the 1:4 position, the proportions of these residues being variable. Alginic acid is hydrophilic and absorbs water, and is soluble in alkaline solutions. The acidic solution (pH 2.0–3.4) gels when treated with a bivalent cation such as Ca^{2+} which forms salt bridges among the carboxylate groups. Chelating agents, such as phosphates, will complex with calcium and damage the gel structure by swelling. Efforts have been made to prevent this effect, mainly because phosphates and related anions occur frequently in biological preparation in buffer solutions, cell metabolites, ATP, etc. Prevention involves using cations of barium [83], strontium [83] and also polycations [113]. Three chemical methods of crosslinking have been devised to give alginate gel resistance to phosphate degradation [114].

Carrageenan. Carrageenan, a polysaccharide extracted from the red seaweed *Eucheumi cottonii*, consists of alternating units of β-D-galactose and 3,3-anhydro-α-D-galactose, of which six types are known depending on the sulphate content of the polymer.

Thus κ-carrageenan forms rigid gels and may shrink; ι-carrageenan forms a nonshrinking flexible gel and λ-carrageenan does not form gels. Carrageenan is soluble in water above 60 °C and in hot concentrated salt solutions of potassium, calcium and sodium and is degraded by acidic solutions (pH < 9). Kappa (κ) carrageenan contains one sulphate for each disaccharide unit and forms a rigid gel under a variety of conditions including cooling, and contact with any one of a large number of metal cations, amines, ammonium ions, and water miscible organic solvents [26].

The most sulphated type, iota-carrageenan, extracted from *Eucheuma spinosa,* forms flexible gels. The lambda carrageenan obtained from *Gigartina aciculaire* and *G. pistillata* do not form gels, but are used to blend with other types to obtain the desired consistency. κ-carrageenan has the advantage over many other gels (naturally occurring polysaccharide gels, synthetic organic gels, inorganic gels) that its conditions for gelation are flexible and varied and enable immobilization to be carried out under the optimum conditions for the operational stability of the microorganisms; the toxic effects of monomers, irradiation, etc., as used in synthetic polymerization are avoided. For this reason, κ-carrageenan has been employed as a carrier in a number of industrial fermentations using immobilized whole cells and enzymes [115]. The conditions under which κ-carrageenan forms a gel have been investigated [116]. Various hardening agents have been used for κ-carrageenan, including glutaraldehyde/hexamethylene-diamine [117, 118], polyethyleneimine [119), and epichlorohydrin/amine [120]. Synthetic organic polymers such as polyacrylamide and the mineral hydroxapatite can be incorporated into the gel structure of κ-carrageenan for increased mechanical strength [11, 84].

122

Cellulose. Compared to polysaccharide gels, cellulose and its acetates have high mechanical strength.

Cellulose is often used by itself or derivatized as an ion-exchanger in immobilization work. Examples include the anion exchanger cellulose derivatives DEAE cellulose (diethylamino ethyl) and ECTEOLA (epichlorhydrin triethanolamine) cellulose. Other cellulose-based anion exchangers include PEI (polyethyleneimine) cellulose, TEAE (triethylaminoethyl)cellulose and QAE (diethyl-[2-hydroxypropyl]-amino ethyl) cellulose. Cellulose di- or triacetate and their mixtures with N-ethylpyridinium chloride are soluble in dimethyl sulphoxide. From such a solution it is possible to form beads by dispersion [46].

Synthetic Organic Carriers. By far the most important synthetic organic carriers used in immobilization work are polymers made from the vinyl carboxylate functional unit,

$$CH_2 = CH - \overset{\overset{\displaystyle O}{\displaystyle \|}}{C} - O-$$

polyacrylamide and substituted polyacrylamides, substituted polyacrylates, and substituted polymethacrylates.

Another functional group, isocyanate (–NCO), is very often used to modify structurally and functionally polymers containing the vinyl carboxylate backbone [121].

Bifunctional monomers such as diisocyanates are the sources of prepolymers for photocrosslinkable polymers and in preparing polyurethanes. Other polymers that are increasingly used are based on polystyrenes, mostly in the form of ion-exchange resins. The use of synthetic organic polymers in cell immobilization may give rise to certain complications in the fermentation process because of the more complex chemical nature of the fermentation mixture. Thus, in addition to the polymers themselves, which may chemically influence the formation, there may be present the unreacted monomer and preservatives added to the monomer in the form of radical scavengers. Polymerization is often initiated by radical chain reaction initiators and some polymerizations are conducted in the presence of one or more crosslinking reagents or copolymers and dispersing agents. All these compounds may be present in the polymer product and may affect the outcome of the fermentation if they are not removed effectively. The usual high viscosity and low porosity of the polymer product contribute to difficulties in purification. Furthermore, most reactive monomers will easily react through oxidation, hydrolysis, or photochemical reactions in contact with air and light, giving products that may be detrimental to the fermentation reaction. Monomer reagents may contain impurities such as metal ions that affect a fermentation, especially if living cells are involved, although such impurities are also known to have deactivating effects on the enzyme contained in non-viable immobilized cells.

Acrylamide, a low melting crystalline solid, has a wide pH range (4.4 to >6) when dissolved in water (5%). It may contain a trace (up to 0.1%) of the free acid and may contain preservatives such as hydroquinone, t-butyl pyrocatechol, and free radical initiators such as peroxides and persulphates. Polyacrylamide hydrolyses in basic aqueous solution to give ammonia as one of the products.

Acrylamide, a vinyl carboxylic acid amide, can polymerize linearly

$$CH_2=CH-\overset{\overset{\displaystyle O}{\|}}{C}-NH_2 \qquad (CH_2-CH)_n$$
$$\qquad\qquad\qquad\qquad\qquad\overset{|}{CONH_2}$$

under irradiation and the catalytic action of a number of reagents such as potassium persulphate or N,N,N',N'-tetramethylenediamine riboflavin. This high molecular weight linear polymer is water miscible and contains a high degree of hydrogen bonding. The polymer turns into a hard solid slab on drying and can be pulverized into granules. Polyacrylamide in solution is susceptible to destruction by shear forces and also to degradation on standing, especially in the presence of ferrous/ferric salts. To improve its mechanical properties, polyacrylamide can be treated with a crosslinking agent. The presence of a bifunctional reagent, e.g., N,N'-methylenebis acrylamide, $(CH_2=CHCONH)_2CH_2$, gives rise to a crosslinked polymer product. Based on the notation

$$-CH_2CH\overset{\overset{\displaystyle O}{\|}}{C}NH_2 = A$$

and

$$(-CH_2\underset{|}{CH}CONH_2)_2CH_2 = B$$

a product such as in Fig. 4.14 is obtained having a lattice-like structure suitable for entrapping microbial cells.

The effects of reagents used in immobilization of cells of *Alcaligenes faecalis* on polyacrylamide have been determined by monitoring the enzyme activity of the β-glucosidase of this microorganism [122]. A loss of β-glucosidase activity has been observed in cells in contact with a mixture of potassium persulphate and β-dimethylaminoproprionitrile, reagents used as polymerization initiators. The opposite effect has been observed with the monomer acrylamide and the crosslinking agent N,N'-methylenebisacrylamide; the activity increased at a rate even greater than that of the cells circulated in a phosphate buffer in the absence of these reagents. This effect is attributed to a possible permeabilizing or even lysis action of these reagents on the cell wall [122]. Such effects of increased activity

124

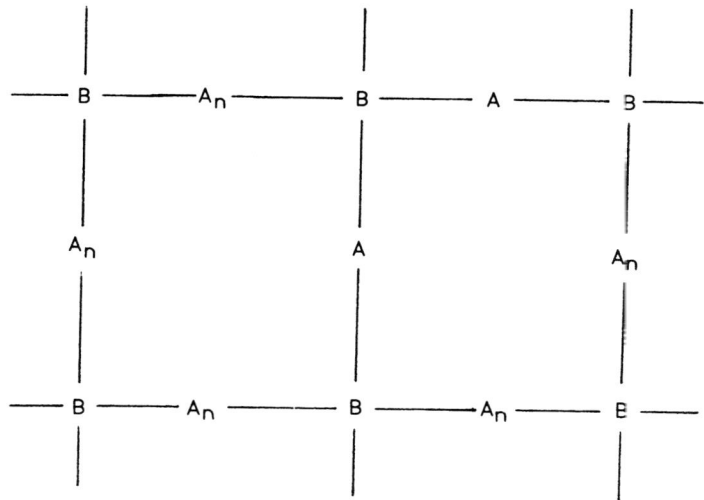

Fig. 4.14. Example of a cross-sectional lattice structure for encapsulating cells

$$A = -CH_2\overset{|}{C}H\overset{O}{\overset{\|}{C}}NH_2$$

$$B = (-CH_2\overset{|}{C}HCONH_2)CH_2$$

due to permeabilization or lysis due to acrylamide have been observed in other systems [122, 123], and a drastic decline in the viability of *Escherichia coli B* cells has been traced to the monomer acrylamide, especially in concentrations of $> 5\%$ [124]. The loss of enzyme activity of *Nocardia erythropolis* after immobilization on polyacrylamide gel is attributed to the toxic effect of the monomer [125]. The generally adverse effects of polyacrylamide [126] as a carrier for the immobilization of cells of *Candida lipolytica* [127, 128] in the production of citric acid via cell leakage and also in immobilization of methane-oxidizing cells [129] perhaps have similar chemical explanations. Chemical effects on the physiology of attached cells immobilized on polyacrylamide were probed in the system of polyacrylamide beads derivatized with primary or tertiary amines [130].

Photocrosslinkable Prepolymers and Polyurethanes. A series of acrylate derivatives was introduced by FUKUI et al. [131] as photocrosslinkable prepolymers for the entrapping of cells and organelles. These prepolymers consist of the water soluble poly(ethylene glycol) methacrylate (PEGM)©, several condensation polymers formed between a hydroxyalkylacrylate

$$CH_2 = CH - \overset{O}{\overset{\|}{C}} - OCH_2CH_2OH$$

and a linear prepolymer with terminal isocyanate groups. These prepolymers are formed between a polyethyleneglycol of chosen chain length and the bifunctional compound isophorone diisocyanate

Polyurethanes contain the repeating carbamate group,

$$-HN-\overset{\overset{\displaystyle O}{\|}}{C}-O-$$

which is formed between a diisocyanate, $R(NCO)_2$, and a glycol unit $HO-[CH_2CH_2O]_nH$.

Bifunctional monomers such as ethylene glycol or other macroglycols of polyethenes or polyesters polymerize to linear polymers with good physical properties but low thermal stability. By the use of higher functional monomers or by treating the polymerizing mixture with water, crosslinked products are produced. In the latter method, the water reacts with the isocyanate group to give CO_2 and an amino group, which can further add to other isocyanate groups to form diuret linkages. The CO_2 gas produced expands and gives the product a foam structure. The chemistry of isocyanate colymers used in cell and enzyme immobilization work has been reviewed by GOLDSTEIN [121]. The advantage of this system of polymers is that structural parameters can be varied in the choice of the prepolymer. Thus the use of polypropylene glycol $HO[CHCH_3CH_2-O]_nH$ instead of polyethyleneglycol changes the polymer from hydrophilic to lipophilic. The chain length of the central polyethyleneglycol unit can be varied to suit the geometrical requirements of enzymes or cells to be immobilized.

The important roles of the ionic nature of hydrophilic and lipophilic carriers in conjunction with different solvent systems in fermentations involving water-insoluble compound have been reviewed by FUKUI and TANAKA [132] and SONOMOTO [133]. By using hydrophobic methane gels as carriers, fermentation of water-insoluble substrates can be conducted in a non-polar organic sorbent with an improved yield [134] over the normal aqueous solution fermentation. Judicious choice of the ionic nature of both the carrier and the solvent system can selectively determine the course of a fermentation transformation [135–137]. Solvent systems commonly used include a water/water-miscible organic solvent [138–140], water-saturated non-polar organic solvent [141] and mixtures of non-polar organic solvents [142].

126

Components in the preparation of urethane prepolymer have been shown to cause a 35-fold increase of β-galactosidase activity in cells of *Caldariella acidophila* [143], an effect very often observed with the reagents involved in the preparation of polyacrylamide. This effect was attributed to the permeabilizing action of those reagents on the cell membrane.

References

1. Bergey's Manual of Determinative Bacteriology (1974[8]) In: Buchanan RE, Gibbons NE (eds). Williams and Wilkins, Baltimore
2. Bushell ME (1983) In: Wiseman A (ed) Principles of biotechnology. Surrey University Press, New York, p 5
3. Navarro JM, Durand G (1980) CR Hebd Seances Acad Sci Ser D 290(6):453
4. Russell RM, Tanner RD (1978) Ind Eng Chem Proc Des Dev 17:157
5. Thomas TD, Batt RD (1969) J Gen Micr 58:371
6. Foerberg C, Enfors SO, Haeggstroem L (1983) Eur J Appl Microbiol Biotechnol 17:143
7. Mosbach K, Birnbaum S, Hardy K, Davies J, Buelow L (1983) Nature (London) 302(5908):543
8. D'Souza SF, Nadkarni GB (1980) Biotechnol Bioeng 22:2191
9. Bandyopadhyay KK, Ghose TK (1982) Bitoechnol Bioeng 24:805
10. Iordan EP, Ikonnikov NP, Kovrizhnykh VA, Vorob'eva LL (1979) Prikl Biokhim Mikrobiol 15:515
11. Kuu WY, Polack JA (1983) Biotechnol Bioeng 25:1995
12. Koshcheenko KA, Sukhodol'skaya GV, Tyurin VS, Skryabin GK (1981) Eur J Appl Microbiol Biotechnol 12:161
13. Wagner F, Lang S (1979) DECHEMA Monograph (1724–1731) 84:315
14. Venkatasubramanian K, Vieth WR (1979) Prog Industrial Microbiol 15:61
15. Klein J, Kressdorf B (1982) Biotechnol Lett 4:375
16. Takara Shuzo Co, Ltd Jpn Kokai Tokkyo Koho (1983) JP 58 23,787 [83 23,787] (Cl. C12N11/04), 12 February; (1982) Appl 82/120,099, 9 July
17. Constantinides A (1980) Biotechnol Bioeng 22:119
18. McGhee JE, St Julian G, Detroy RW, Bothast RJ (1982) Biotechnol Bioeng 24:1155
19. Wagner F, Klein J (1979) DECHEMA 84:265
20. Lamanna C, Mallette MF, Zimmerman LN (1973[4]) Basic bacteriology, its biological and chemical background. Williams and Wilkins, Baltimore, p 744
21. Stieber RW, Gerhardt P (1981) Biotechnol Bioeng 23:535
22. Cheetham PSJ, Imber CE, Isherwood J (1982) Nature (London) 299:628
23. Costerton JW, Marks I (1977) Electron microscopy of enzymes. Principles and methods, vol 5, p 99
24. Fukui S, Kawamoto S, Yashuhara S, Tanaka A (1975) Eur J Biochem 59:561
25. Wada M, Kato J, Chibata I (1980) J Ferment Technol 58:327
26. Kokubu T, Karube I, Suzuki S (1978) Eur J Appl Microbiol 5:233
27. Chibata I (1979) Food Proc Engineering 2:1
28. Sivaraman H, Seetarama Rao B, Pubdle AV, Sivaraman C (1982) Biotechnol Lett 4:359
29. Sommerville HJ, Mason JR, Ruffell RN (1977) Eur J Appl Microbiol 4:75
30. Wheatley MA, Phillips CR (1984) Biotechnol Bioeng 26:583
31. Malmqvist M, Hofsten BV (1975) J Gen Microbiol 87:167
32. Mohan RR, Li NN (1975) Biotechnol Bioeng 17:1137
33. Gaden EL Jr (1981) Scientific American 245:184, 186, 191, 194, 196
34. Brooker BE, Fuller R (1976) In: Fuller R, Lovelock DW (eds) Microbial ultrastructure, the use of the electron microscope. Academic Press, New York, p 87
35. Fletcher M, Floodgate GD (1976) In: Fuller R, Lovelock DW (eds) Microbial ultrastructure, the use of the electron microscope. Academic Press, New York, p 101

36. Casida LE Jr (1968) Industrial microbiology. Wiley, New York, p 164
37. Mosbach K (1983) Trans R Soc Lond B300:355
38. Erickson RJ (1976) Microbiology (Washington, DC), p 406
39. Chibata I (ed) (1978) Immobilized enzymes. In: Research and development. Wiley, New York, p 136
40. Chibata I (ed) (1978) Immobilized enzymes. In: Research and development. Wiley, New York, p 139
41. Chibata I, Tosa T (1977) Adv Appl Microbiol 22:1
42. Chibata I (ed) (1978) Immobilized enzymes. In: Research and development. Wiley, New York, p 124
43. Chibata I (ed) (1978) Immobilized enzymes. In: Research and development. Wiley, New York, p 122
44. Cheetham PSJ (1980) Topics in enzyme. Ferment Biotechnol 4:189
45. Venkatasubramanian K (1979) Food Proc Eng 2:162
46. Linko P, Poutanen K, Weckstrom L, Linko Y-Y (1980) Biochemie 62:387
47. Hornby WE, Lilly MD, Crook M (1968) Biochem J 107:669
48. Schuler ML, Aris R, Tsuchiya HM (1972) J Theor Biol 35:67
49. Goldman R, Kedem O, Katchalski E (1968) Biochem 7:4518
50. Marsh DR, Lee YY, Tsao GT (1973) Biotechnol Bioeng 15:483
51. Goldstein L (1976) In: Mosbach K (ed) Methods of enzymology. 44:397
52. Messing RA (1978) Adv Biochem Eng 10:51
53. Buchholz K (ed) (1979) Characterization of immobilized biocatalysis. In: DECHEMA monographs, vol 84. Verlag Chemie, New York, pp 1724–1731
54. Linko YY, Pohjola L, Linko P (1977) Proc Biochem 12:14
55. Lasch J, Iwig M, Koelsch R (1975) Eur J Biochem 60:163
56. Chen LF, Tsao GT (1976) Biotechnol Bioeng 18:1507
57. Amsterdam A, Er-El Z, Shaltiel S (1975) Arch Biochem Biophys 171:673
58. Schlünsen J, Ehrenthal E, Manecke G (1979) In: Buchholz K (ed) Characterization of immobilized biocatalysts. Verlag Chemie, New York
59. Rüchel R, Brager MD (1975) Anal Biochem 68:415
60. Barbotin JN, Thomasset B (1980) Biochemie 62:359
61. Allen T (1975[2]) Particle size measurement. Chapman and Hall, London
62. Buchholz K (ed) (1979) Characterization of immobilized biocatalysts. Verlag Chemie, New York, p 111
63. Schlünsen J (1979) In: Buchholz K (ed) Characterization of immobilized biocatalysts. Verlag Chemie, New York, p 118
64. Buchholz K, Gödelmann B (1979) In: Buchholz K (ed) Characterization of immobilized biocalysts. Verlag Chemie, New York, p 127
65. Ziomek E, Martin WG, Veliky IA, Williams RE (1982) Enzyme Microbiol Technol 4(6):405
66. Allen T (1975[2]) Particle size measurement. Chapman and Hall, London, p 76
67. Atkinson B, Black GM, Pinches A (1980) Proc Biochem 15:24
68. Blieva RK (1982) Mikrobiologiya 51:945
69. Allen T (1975[2]) Particle size measurement. Chapman and Hall, London, p 414
70. Jelinck ZK (1974) Particle size analysis. Wiley, New York, p 126
71. Miyawaki O, Nakamura K, Tano T (1980) Agric Biol Chem 44:2865
72. Messing RA (1974) Research/Development 25:32
73. Messing RA (ed) (1975) Immobilized engineering for industrial reactions. Academic Press, New York
74. Haller W (1983) In: Scouten WH (ed) Solid state biochemistry, analytical and synthetic aspects. Wiley, New York, p 535
75. Messing RA, Oppermann RA, Kolot FB (1979) In: Venkatasubramanian K (ed) Immobilized microbial cells. ACS Symposium Series 106, Washington, p 13
76. Messing RA (1980) Ann Rep Ferment Proc 4:105
77. Kumakura M, Yoshida M, Kaetsu I (1979) Appl Environ Microbiol 37:310
78. Messing RA, Oppermann RA, Ramsey WS, Takeguchi MM (1981) US Patent No 4,246,349, 20 January

79. Navarro AR, Lucca ME, Callieri DAS (1982) Acta Cient Venez 33(3):214
80. Parascandola P, Salvadore S, Scardi V (1982) J Ferment Technol 60:477
81. Tso WW (1980) Biotechnol Lett 2:519
82. Van Suijdam JC, Kossen NWF, Paul PG (1980) Eur J Appl Microbiol Biotechnol 10(3):211
83. Paul F, Vignais PM (1980) Enzyme Microbiol Technol 2:281
84. Wang HY, Hettwer D (1982) Biotechnol Bioeng 24:1827
85. Brauer H, Mewes D (1972) Chem Ing Techn 44:93
86. Ergun S (1952) Chem Eng Progr 48:89
87. Buchholz K, Gödelmann B (1979) Enzyme Engineering 4:89
88. Tiller FM (1975) Filtration Separation 4:386
89. Klein J, Kluge M (1979) DECHEMA Monographs 84:285
90. Suhaila M, Salleh AB (1982) Biotechnol Bioeng 4:611
91. Cheetham PSJ, Blund KW, Bucke C (1979) Biotechnol Bioeng 21:2155
92. Zueva NN, Shchrbakoba VN, Yakovlea VYa, Nikitin YuS, Avsyuk IV, Chan Tkhi Tuet Mai, Berezin IV (1980) Prikl Biokhim Microbiol 16:918
93. Buchholz K (ed) (1979) DECHEMA Monographs 84:136
94. Klein J, Eng H (1979) DECHEMA Monographs 84:292
95. Wang HY, Lee SS, Takach Y, Cawthon L (1982) Biotechnol Bioeng Symp 12:139
96. Rouxhet PG, Van Haecht JL, Didelez J, Gerard P, Briquet M (1981) Enzyme Microb Technol 3:49
97. Krouwel PG, Kossen NWF (1980) Biotechnol Bioeng 22:681
98. Gerson DF, Zajic JE (1979) In: Venkatasubramanian K (ed) Immobilized microbial cells. ACS Symposium Series 106. American Chemical Society, Washington DC, p 29
99. Ellwood DC, Melling J, Rutter P (eds) (1979) Adhesion of microorganisms to surfaces. Academic Press, New York
100. Lips A, Jessup NE (1979) In: Ellwood DC, Melling J, Rutter P (eds) Adhesion of microorganisms to surfaces. Academic Press, New York, p 5
101. Rogers HJ (1979) In: Ellwood DC, Melling J, Rutter P (eds) Adhesion of microorganisms to surfaces. Academic Press, New York, p 29
102. Thonart P, Custinne M, Paquot M (1982) Enzyme Microbiol Technol 4:191
103. Daniels SL (1971) Dev Ind Microbiol 13:211
104. Marshall KC (1976) Interfaces in microbial ecology. Harvard Univ Press, Cambridge
105. Atkinson B, Daould IS (1976) Adv Biochem Eng 4:42
106. Bories A, Raynal J, Maugenet J (1978) J Ind Aliment Agric 1103
107. Thonart P, Paquot M, Baijot B, Michaux M, Deroanne C (1982) Belg Patent No BE 890,811, 15 February
108. Manecke G, Ehrenthal E, Schlünsen J (1979) DECHEMA Monographs 84(1724–1731):49
109. Ramesh V, Singh C (1981) Enzyme Microb Technol 3:246
110. Navarro JM, Durand G (1981) Ann Microbiol 132B:241
111. Tischer W, Tiermeyer W, Simon H (1980) Biochimie 62:331
112. Yamada H, Yamada K, Kumagoi H, Hino T, Okamura S (1978) Enzyme Eng 3:57
113. Veliky LA, Williams RE (1981) Biotechnol Lett 3:275
114. Birnbaum S, Pendleton R, Larsson PO, Mosbach K (1981) Biotechnol Lett 3:393
115. Chibata I (1979) In: Venkatasubramanian K (ed) Immobilized microbial cells. ACS Symposium Series 106. Am Chem Soc, Washington DC, p 187
116. Tosa T, Sata T, Mori T, Yamamoto K, Takata I, Nishida Y, Chibata I (1979) Biotechnol Bioeng 21:1697
117. Murata K, Tani K, Kato J, Chibata I (1980) Eur J Appl Microbiol Biotechnol 10:11
118. Nishida Y, Sato T, Tosa T, Chibata I (1979) Enzyme Microbiol Technol 1:95
119. Takata I, Kayashima K, Tosa T, Chibata I (1982) J Ferment Technol 60:431
120. Takata I, Kayashima K, Tosa T, Chibata I (1982) J Appl Biochem 4:371
121. Goldstein L (1980) Biochem 62:401
122. Wheatley MA, Phillips CR (1981) Adv Biotechnol 2:47
123. Martin CKA, Perlman D (1976) Biotechnol Bioeng 18:217
124. Starostina NG, Lusta KA, Fikhte BA (1982) Prikl Biokhim Microbiol 18:225

125. Atrat P, Huller E, Horhoid C, Buchar MJ, Arinbasarova AY, Koschtschejenko KA (1980) Z Allg. Mikrobiol 20:159
126. Freeman A, Aharonowitz Y (1981) Biotechnol Bioeng 23:2747
127. Stottmeister U (1979) Z Allg Mikrobiol 19:763
128. Berger R, Langhammer G (1980) Z Allg Mikrobiol 20:69
129. Romanovskaya VA, Karpenko VI, Pantskhava ES, Grinberg TA, Malashenko YR (1981) Adv Biotechnol [Proc Int Ferment Symp 6th 1980] 3:367
130. Renveny S, Mizrahi A, Kotler M, Freeman A (1983) Biotechnol Bioeng 25(2):469
131. Fukui S, Sonomoto K, Itoh S, Tanaka A (1980) Biochimie 62:381
132. Fukui S, Tanaka A (1982) Ann Rev Microbiol 36:145
133. Sonomoto K, Usui N, Tanaka A, Fukui S (1983) Eur J Appl Microbiol Biotechnol 17:203
134. Omata T, Iida T, Tanaka A, Fukui S (1979) Eur J Appl Microbiol Biotechnol 8:143
135. Fukui S, Ahmed SA, Omata T, Tanaka A (1980) Eur J Appl Microbiol Biotechnol 10:289
136. Omata T, Tanaka A, Fukui S (1980) J Ferment Technol 58:339
137. Fukui S, Tanaka A, Gellf G (1978) Enzyme Eng 4:299
138. Fukui S, Yokozeki K, Yamanaka S, Ulapawa T, Takinami K, Hirose Y, Tanaka A, Sonomoto K (1982) Eur J Appl Microbiol Biotechnol 14:225
139. Sonomoto K, Hoq M, Tanaka A, Fukui S (1983) Appl Environ Microbiol 45:436
140. Sonomoto K, Tanaka A, Omata T (1979) Eur J Appl Microbiol Biotechnol 6:325
141. Omata T, Iwamoto N, Kimura T, Tanaka A (1981) Eur J Appl Microbiol Biotechnol 11:199
142. Yamane T, Nakatani H, Sada E, Omata T, Tanaka T, Fukui S (1979) Biotechnol Bioeng 21:2133
143. Drioli E, Iorio G, Santoro R, De Rosa M, Gambacorta A, Nicolaus B (1982) J Mol Catal 14:247
144. Seyhan F, Kirwan DJ (1979) Biotechnol Bioeng 21(2):271
145. Chang Y-Y, Wang C-H, Wang H-C, Chang C-H (1980) Wei Sheng Wu Hsueh T'ung Pao 7:112
146. Chua JW, Eraeslan A, Kinoshita S (1980) J Ferment Technol 53:123
147. Grote W, Lee KJ, Rogers PL (1980) Biotechnol Lett 2:481
148. Morikawa Y, Karube I, Suzuki S (1980) Biotechnol Bioeng 22:1015
149. Yang L-W, Zhong L-C (1980) Wei Sheng Wu Hsueh Pao 20:296
150. Yamamoto K, Tosa T, Chibata I (1980) Biotechnol Bioeng 22:2045
151. Haeggstroem L (1981) Adv Biotechnol [Proc Int Ferment Symp, 6th 1980] (Pub 1981) 2:79
152. Margaritis A, Bajpai PK, Wallace JB (1981) Biotechnol Lett 3:613
153. Constantinides A, Bhatia D, Vieth WR (1981) Biotechnol Bioeng 23:899
154. Karube I, Kokubu T, Suzuki S (1981) Biotechnol Bioeng 23:29
155. Linko P, Poutanen K, Lino Y-Y (1981) J Mol Catal 13:263
156. Linko Y-Y, Jalanka H, Linko P (1981) Biotechnol Lett 3:263
157. Veliky IA, Jones A (1981) Biotechnol Lett 3:551
158. Weetall HH, Sharma BP, Detar CC (1981) Biotechnol Bioeng 23:605
159. Wada M, Kato J, Chibata I (1981) Eur J Appl Microbiol Biotechnol 11:67
160. Kokufuta E, Matsumoto W, Nakamura I (1982) Biotechnol Bioeng 24:1591
161. Kim HS, Dewey DYR (1982) Biotechnol Bioeng 24:2167
162. Kluge M, Klein J, Wagner F (1982) Biotechnol Lett 4:293
163. Linko P, Stenroos S, Linko Y (1982) Biotechnol Lett 4:159
164. Veelken M, Pape H (1982) Eur J Appl Microbiol Biotechnol 15:206
165. Spettoli P, Nuti MP, Bottacin A, Zamorani S (1982) Util Enzymes Technol Aliment, Symp Int, p 545
166. Szwajcer E, Brodelius P, Mosbach K (1932) Enzyme Microb Technol 4:409
167. Margaritis A, Bajpai P (1982) Biotechnol Bioeng 24:1483
168. Egerer P, Simon H (1982) Biotechnol Lett 4:501
169. Stenroos SL, Linko YY, Linko P (1982) Biotechnol Lett 4:159
170. Matteau PP, Saddler JN (1982) Biotechnol Lett 4:715
171. Deo YM, Gaucher GM (1983) Biotechnol Lett 5:125

172. Hsiao HY, Chiang LC, Yang CM, Chen LF, Tsao GT (1983) Biotechnol Bioeng 25:363
173. Ferenci T (1983) Appl Environ Microbiol 45:384
174. Bang WG, Behrendt U, Lang S, Wagner F (1983) Biotechnol Bioeng 25:1013
175. Schlünsen J, Ehrenthal E, Manecke G (1979) In: Buchholz K (ed) Characterization of immobilized biocatalysts. Verlag Chemie, New York, p 140
176. Tanaka A, Yasuhara S, Osumi M, Fukui S (1977) Eur J Biochem 80:193
177. Krämer DM (1979) DECHEMA Monographs 84:88
178. Garde L, Thomasset B, Tanaka A, Gellf G, Thomas D (1981) Eur J Appl Microbiol Biotechnol 11:133

5 Kinetics and Reactor Design for Immobilized Cells

5.1 Introduction

Design procedures for immobilized cell reactors are analogous to those used for solid-catalyzed chemical reactors, with the major difference being that immobilized cell kinetics are used instead of chemical kinetics. The extensive literature for the design of chemical reactors is not reviewed here; for this, the reader should refer to one of the many excellent texts in the area, for example, [1–3]. In this chapter, special considerations for immobilized cell reactors are discussed. Special considerations for immobilized enzyme reactors have been discussed by PITCHER [4].

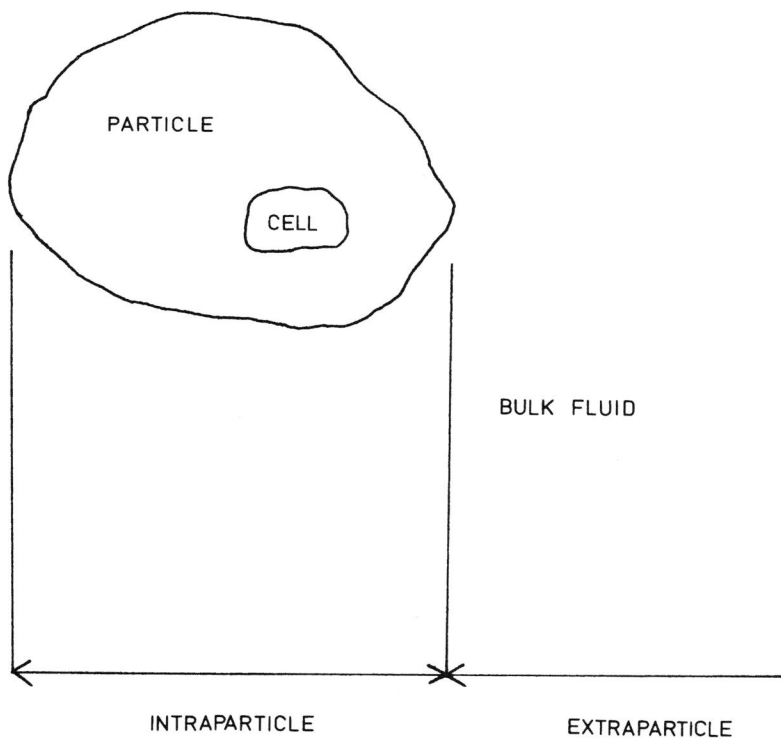

Fig. 5.1. Intra- and extra-particle mass transfer resistances

The fundamental equation for immobilized cell kinetics can usually be taken to be Michaelis or Monod in form. In the absence of complicating factors, and in the absence of mass transfer resistance, Michaelis-type kinetics may describe the system adequately. Mass transfer is important in many instances, however, particularly for encapsulated cells, in which case both reactants and products must diffuse between the cell and the medium. In some cases, mass transfer may control.

It is convenient to classify mass transfer as either extraparticle (that is, concerning the medium external to the particle containing the cell) or intraparticle (that is, within the particle containing the cell). Extraparticle mass transfer is extensively studied in chemical engineering, and the body of literature is vast [5]. The relation between the two forms of mass transfer is illustrated diagrammatically in Fig. 5.1.

The species being transformed may exist in solution in the liquid phase or as a separate gas phase (e.g. oxygen, carbon dioxide) or solid phase (e.g., cellulose). There are many examples of the effects of intraparticle diffusion in the literature [6–8].

5.2 Effectiveness Factors

Since most of the reaction takes place in the interior of the particle, substrate must move through the pores into the particle and products must diffuse out of the particle. The observed rate of reaction will differ from the intrinsic rate of reaction because of resistance to this pore diffusion. The effectiveness factor is a measure of the combined effect of resistance to pore diffusion and resistance to reaction.

The effectiveness factor, η, of an immobilized cell system can be defined as [9]

$$\eta = \frac{v_{immob}}{v_{free}},$$

where

$$v_{immob} = \text{rate of reaction of immobilized cells},$$

$$v_{free} = \text{rate of reaction of free cells}.$$

The effectiveness factor can be expressed as a function of three parameters, the Sherwood number, Sh, the Thiele modulus, ϕ, and the substrate concentration, S:

$$\eta = f(Sh, \phi, S), \tag{1}$$

where

$$Sh = \text{Sherwood number}$$

$$= \frac{k_s d_p}{D},$$

where

k_s = mass transfer coefficient,

d_p = particle diameter,

D = diffusion coefficient of controlling species.

The Sherwood number is the ratio of extraparticle convective mass transfer to molecular mass transfer. It is also the ratio of the concentration gradient at the particle surface to the overall external concentration gradient. If the Sherwood number is large, the difference between the substrate concentration in the bulk phase, S_L, and the substrate concentration in the surface region of the particle, S_R, is small, and extraparticle mass transfer is not rate-limiting.

The Thiele modulus is a measure of the relative effect of intraparticle reaction versus diffusion; for large values of the Thiele modulus, the diffusional resistance is high, and for small values of the Thiele modulus, the diffusional resistance is low.

Reaction occurs on the pore walls where cells are immobilized. Intraparticle diffusive resistance results in a concentration profile within the particle since substrate must diffuse from the bulk fluid. The average reaction rate within the pore is therefore less than if the substrate concentration in the pore were everywhere the same as in the surface region of the particle.

The mass balance for reaction, extraparticle mass transfer and intraparticle diffusion can be written [9] as

$$v_c = k_s \left(\frac{A}{V_c}\right)(S_L - S_R) = D_e\left(\frac{A}{V_c}\right)\left(\frac{dS}{dr}\right)_R, \tag{2}$$

where

v_c = rate of reaction based on a unit volume of particle containing immobilized cells,

A = surface area of particle containing immobilized cells,

V_c = volume of particle,

S_L = substrate concentration in the bulk phase,

S_R = substrate concentration in the surface region of the particle,

D_e = effective diffusion coefficient in the particle, and is equal to D(porosity/tortuosity). Porosity is the void fraction, tortuosity is the ratio of the actual diffusion path to the straight line distance,

r = radial coordinate,

R = radius of particle.

135

The equation for steady state diffusion and reaction in a spherical particle is

$$\frac{v_r}{D_e} = \frac{d^2S}{dr^2} + \frac{2}{r}\frac{dS}{dr},$$ (3)

where

S = substrate concentration at radius r,

v_r = reaction rate at radius r.

If the reaction is first order and irreversible, the reaction rate at radius r is given by

$$v_r = k_1 S,$$ (4)

where
k_1 = intrinsic first order reaction rate constant
and the overall reaction rate for unit volume of porous carrier is

$$v_c = k_1' S_R = \eta k_1 S_R,$$ (5)

where
k_1' is the observed reaction rate constant.

The effectiveness factor, η, previously defined as the ratio of the rate of reaction of immobilized cells to the rate of reaction of free cells, may now be defined in terms of the immobilized system only as

$$\eta = \frac{\text{actual reaction rate within pore}}{\text{reaction rate assumed if } S = S_R \text{ everywhere}}.$$

This effectiveness factor takes into account only the effect of intraparticle diffusion resistance and is the definition usually used in chemical reaction engineering.

A different form of the effectiveness factor is also widely used in biochemical engineering, namely, the overall effectiveness factor (FINK et al. [10], CHEN et al. [11], and YAMANE et al. [12]). The overall effectiveness factor, η_{ov}, is defined as

$$\eta_{ov} = \frac{\text{actual reaction rate within pore}}{\text{reaction rate assumed if } S = S_L \text{ everywhere}}.$$

The overall effectiveness factor includes the effects of both extraparticle and intraparticle diffusional resistance. It is used mostly in biochemical engineering.

For an irreversible first order reaction, Eq. (3) becomes

$$\frac{k_1 S}{D_e} = \frac{d^2S}{dr^2} + \frac{2}{r}\frac{dS}{dr}.$$ (6)

136

Solution of Eq. (6) to give a concentration profile allows η to be evaluated. The result is

$$\eta = \frac{3}{\phi}\left(\frac{1}{\tanh\phi} - \frac{1}{\phi}\right), \tag{7}$$

where

$\phi =$ the Thiele modulus for irreversible first order chemical kinetics, defined as $\phi = R(k_1/D_e)^{1/2}$.

As an approximation,

$$\eta \to \frac{3}{\phi} \quad \text{for} \quad \phi > 20. \tag{8}$$

For $\phi \to 0$, $\eta \to 1$, which implies that the concentration of substrate does not drop appreciably within the pore, and the average rate of reaction approaches the reaction rate on the surface of the particle. A short pore (small particle), slow reaction, or fast diffusion all contribute to making ϕ small.

A high reaction rate, large particle size, or a small diffusion coefficient all cause ϕ to become large. For large ϕ, the substrate concentration decreases rapidly away from the entrance of the pore, and the overall reaction rate observed is less than the reaction rate on the surface. The magnitude of Thiele modulus, ϕ, therefore indicates whether the effect of diffusion is negligible.

From the discussion above, a first order irreversible chemical reaction with the rate of reaction expressed by $v_r = k_1 S$ leads to the definition of the Thiele modulus

$$\phi = R(k_1/D_e)^{1/2}.$$

For cells, the kinetics equation is Michaelis, namely,

$$v_r = \frac{v_{max}S}{K_m + S}, \tag{9}$$

where

$K_m =$ Michaelis constant,

$V_{max} =$ maximum rate of reaction.

Accordingly, for cells, k_1 is replaced by $\dfrac{V_{max}}{K_m + S}$, and the Thiele modulus (for Michaelis-Menten kinetics) becomes

$$\phi = R\left(\frac{V_{max}}{D_e(K_m + S)}\right)^{1/2}.$$

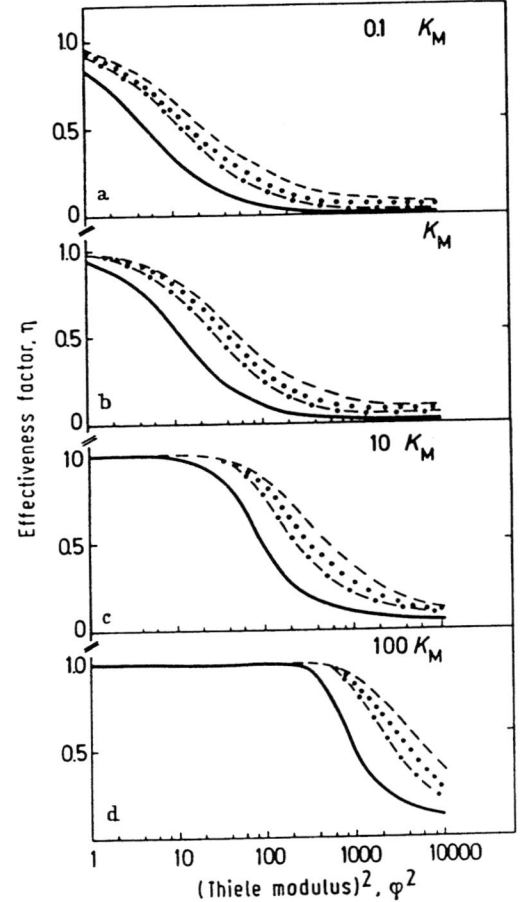

V_{max} per $(K_M P_E)$				D_e (cm^2 s^{-1})	d_p (μm)
0.1	1	10	100	5×10^{-6}	500
1	10	100	1000	5×10^{-6}	100
0.01	0.1	1	10	5×10^{-7}	500
0.1	1	10	100	5×10^{-7}	100

Fig. 5.2 a–d. The stationary effectiveness factor (η) for a one substrate enzyme immobilized in a spherical particle as a function of the square of the Thiele modulus (proportional to the enzyme content), Sherwood number, and initial substrate content in units of K_M. The upper scale gives V_{max}/K_M for different particle dimensions (d_p) and substrate diffusion coefficients (D_e). The corresponding Thiele modulus is given by the lower scale (I). Initial substrate content: **a** 0.1 K_M; **b** K_M; **c** 10 K_M; **d** 100 K_M; ------- Sh′ = 100; .... Sh′ = 16; —·—·— Sh′ = 8; ———— Sh′ = 2. P_E: fraction of particle volume which is enzyme. (Fraction of pore volume in general in the range of 0.3–0.5)

If $S \gg K_m$, the Michaelis-Menten reaction approaches zero order with $v = V_{max}$, and, on replacing k_1 with $\dfrac{V_{max}}{S}$, the Thiele modulus becomes

$$\phi_0 = R\left(\frac{V_{max}}{D_e S}\right)^{1/2}.$$

If $S \ll K_m$, the Michaelis-Menten reaction approaches first order with

$$v = \frac{V_{max}}{K_m} S$$

and the Thiele modulus becomes

$$\phi_1 = R\left(\frac{V_{max}}{D_e K_m}\right)^{1/2}.$$

Michaelis-Menten kinetics lie between zero and first order.

When the intrinsic reaction rate is represented by a nonlinear Michaelis-Menten equation, the effectiveness factor, η, is determined by numerical solution of Eqs. (2) and (3). The observed overall reaction rate for unit volume of porous particle is

$$v_c = \eta \frac{V_{max} S_R}{K_m + S_R}. \tag{10}$$

BUCHHOLZ [9] has provided graphical solutions for parameters of interest in immobilized enzyme applications (Fig. 5.2 a–d):

It should be noted that in Fig. 5.2 the Thiele modulus is defined as $R(V_{max}/(D_e K_m))^{1/2}$, regardless of the relative magnitudes of S and K_m. The stationary effectiveness factor in Fig. 5.2 refers to constant substrate concentrations in the bulk fluid as in continuous stirred tank reactors.

5.3 Reactor Types and Immobilized Cell Geometries

In batch reactors, the bulk substrate concentration varies with time and in continuous plug flow reactors the concentration changes along the reactor. The stationary effectiveness factor therefore varies with conversion. For these cases, an operational effectiveness factor is used. This effectiveness factor compares the times required for a given degree of substrate conversion (based on activity) by free and by immobilized cells. Graphical correlations for 90% conversion are presented in Fig. 5.3. The definition of the Thiele modulus is $R(V_{max}/(D_e K_m))^{1/2}$, as in Fig. 5.2.

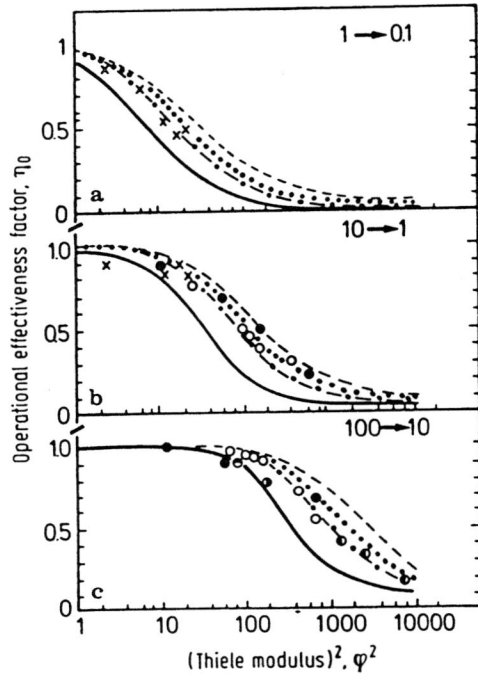

V_{max} per $(K_M P_E)$				D_e (cm² s⁻¹)	d_p (μm)
0.1	1	10	100	5×10^{-6}	500
1	10	100	1000	5×10^{-6}	100
0.01	0.1	1	10	5×10^{-7}	500
0.1	1	10	100	5×10^{-7}	100

Fig. 5.3a–c. Calculated and experimental data for the operational effectiveness factor (η_{op}) for 90% substrate conversion, with a one substrate enzyme immobilized in spherical particles, as a function of the square of the Thiele modulus (proportional to the enzyme content), Sherwood number and initial substrate concentration, given in K_M-units, in the upper right hand corner (1). For the relation between the upper and the lower scale, see legend to Fig. 5.2. Curves: Calculated data for ----- Sh′ = 100; ····· Sh′ = 16; ---- Sh′ = 8; —— Sh′ = 2. Experimental data: × α-chymotrypsin bound to Sepharose 4B; ○ trypsin bound to Sepharose 4B; ● trypsin bound to Sepharose CL 2B; ◐ trypsin bound to isothiocyanatostyrol acrylic acid matrix; ◑ trypsin bound to porous glass Servachrom G 550; ◒ trypsin bound to Oxiran 5120 B

BISCHOFF [13] has expressed the effectiveness factor η as a function of a general modulus M_m. For a flat plate geometry, M_m is given by

$$M_m = L \left(\frac{V_{max}}{2K_m D_e} \right)^{1/2} \left(\frac{S}{K_m + S} \right) \left(\frac{S}{K_m} - \ln(1 + S/K_m) \right)^{-1/2}. \tag{11}$$

where L is the plate thickness. Figure 5.4 shows η as a function of M_m for several values of S/K_m (4). For $M_m > 2$, $\eta = 1/M_m$.

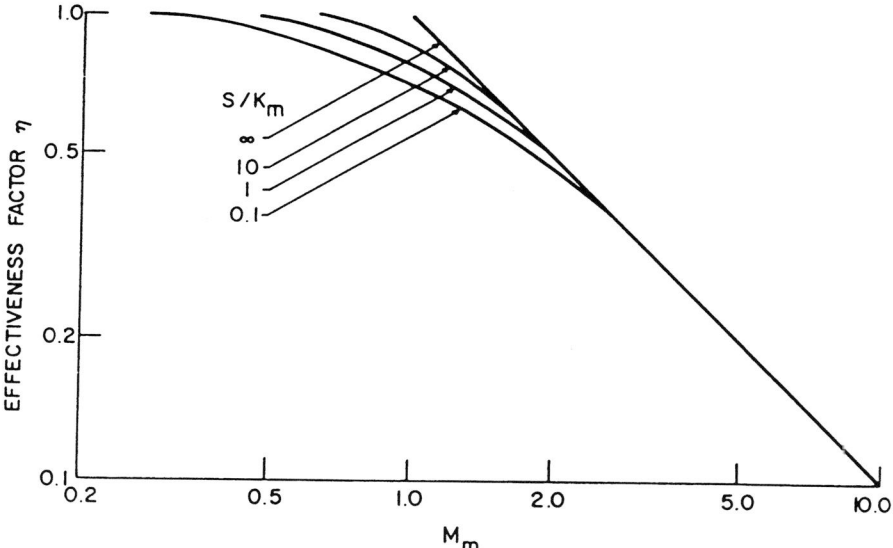

Fig. 5.4. Effectiveness factor, η, vs.

$$M_m = L \left(\frac{V_{max}}{2K_m D_e} \right)^{1/2} \left(\frac{S}{K_m + S} \right) \left[\frac{S}{K_m} - \ln \left(1 + \frac{S}{K_m} \right) \right]^{-1/2}$$

for Michaelis-Menten kinetics (4)

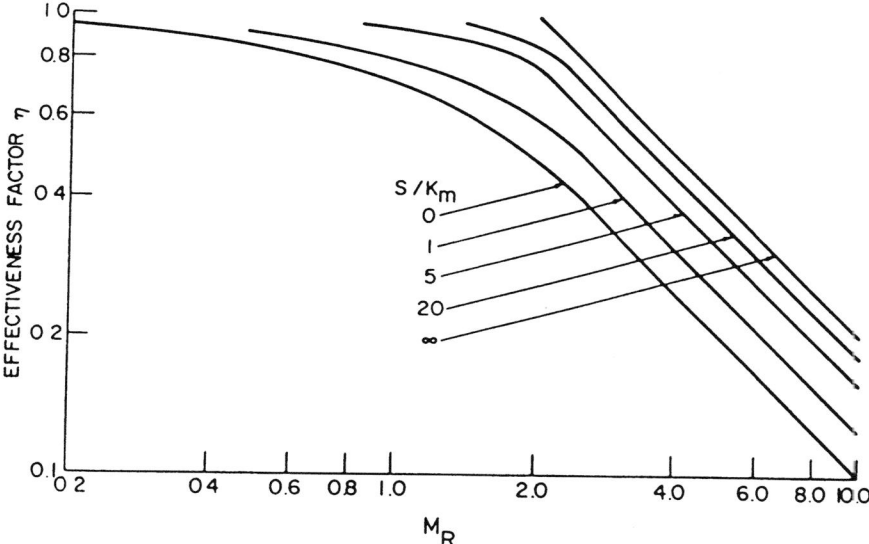

Fig. 5.5. Effectiveness factor, η, vs.

$$M_R = \frac{L^2}{D_e} \left(\frac{1}{V_c} \frac{dS}{dt} \right) \frac{1}{S}$$

for Michaelis-Menten kinetics, flat plate geometry [4]

If observed reaction rates are used instead of a Michaelis kinetic model, a rate-based modulus M_R can be used. For a flat plate geometry,

$$M_R = \frac{L^2}{D_e}\left(\frac{1}{V_c}\frac{dS}{dt}\right)\frac{1}{S}, \tag{12}$$

where $\frac{1}{V_c}\frac{dS}{dt}$ is the reaction rate for unit volume of porous carrier. A plot of η as a function of M_R for several values of S/K_m (Fig. 5.5) is given by PITCHER [4].

If the substrate is partitioned between the immobilizing medium and the bulk liquid phase, it is possible to include the partition coefficient, K_p, defined as the ratio of the concentration of the substrate in the medium volume (pores plus immobilizing matrix) to that in the outside liquid, in one of the boundary conditions so that the resulting value of the overall effectiveness factor incorporates the effect of partition explicitly [12]. The overall effectiveness factor, η_{ov}, as defined earlier, includes the effect of extraparticle mass transfer. Although the decreased free volume available to the substrate in the gel by virtue of the presence of gel material and the immobilized cells might lead to values of K_p less than unity, physico-chemical interactions between substrate and gel material often result in values which are greater than unity, as illustrated in Table 5.1 [12].

YAMANE et al. [12] have derived analytical expressions for the overall effectiveness factor for irreversible zero- and first-order kinetics for slab, rod, and spherical geometries. Numerical solutions have been computed for irreversible Michaelis-Menten kinetics and for substrate inhibition in a spherical geometry. For a first order reaction and the Biot number, $Bi = \frac{k_s R}{D_e} \to \infty$, the effectiveness factor, η, when $K_p \neq 1$ can be obtained simply by multiplying η when

Table 5.1. Partition coefficients for some immobilized enzyme systems [12]

Substrate	Polymer support	K_p	Enzyme	Ref.
Benzoyl L-arginine amide	Polyanionic gel (co-polymer of maleic acid and ethylene)	34 ± 10 (at low ionic strength) 1.32 ± 0.3 (at high ionic strength)	Trypsin	[14]
α-Benzoyl L-arginine ethyl ester	CM-cellulose	$2 \sim 16$	Bromelain	[15]
o-Nitrophenyl β-D-galactopyranoside	Polyacrylamide (7.5%) Polyacrylamide (15%)	1.5 0.89 ± 0.01	β-Galac-tosidase	[16]
o-Nitrophenyl β-D-galactopyranoside	Poly(2-hydroxyethyl methacrylate)	1.29	β-Galac-tosidase	[17]
Hydrocortisone	Photo-crosslinkable resin	3.21	Steroid Δ^1-de-hydrog-enase	[18]

142

Table 5.2. Analytical solution of η for first- and zero-order reactions in spherical geometry [12]

First order

$$\frac{1}{\eta_1} = \frac{\phi_1^2}{3K_p(\phi_1 \coth \phi_1 - 1)} + \frac{\phi_1^2}{3Bi}$$

Zero order

when $\quad \phi_0 \geqq \sqrt{\dfrac{6}{1/K_p + 2/Bi}}, \quad \dfrac{1}{\phi_0^2} - \dfrac{\eta_0}{3Bi} - \dfrac{1}{6K_p}\{1 - (1-\eta_0)^{1/3}\}^2\{2(1-\eta_0)^{1/3} + 1\}$ *

when $\quad \phi_0 \leqq \sqrt{\dfrac{6}{1/K_p + 2/Bi}}, \quad \eta_0 = 1$

* if $K_p/Bi < 3/2$ and $K_p/Bi \neq 1$,

$$\eta_0 = 1 - \frac{1/2 + \cos(\theta/3 + 4\pi/3)^2}{1 - K_p/Bi},$$

where $\quad \theta = \cos^{-1}\{1 - 4(1 - K_p/Bi)^2(K_p/Bi + 1/2 - 3K_p\phi_0^2)\}$

* and if $K_p/Bi = 1$,

$$\eta_0 = 1 - (1 - 2K_p/\phi_0^2)^{3/2}$$

$K_p = 1$ by K_p (see Table 5.2). In the reaction-limited region, that is, $\eta = 1$, K_p has no effect on zero-order reactions, regardless of the magnitude of K_p.

There have been many attempts to derive approximate expressions for the effectiveness factor for immobilized cells with Michaelis-Menten kinetics [19–23]. Numerical solutions are readily calculable [23, 12].

Solutions calculated by YAMANE [12] for irreversible Michaelis-Menten kinetics are shown graphically in Figs. 5.6 and 5.7 [12]. In Figs. 5.6 and 5.7 the parameter κ is defined as K_m/S_L.

For substrate inhibition, a calculated solution is shown in Fig. 5.8 [12].

For immobilized enzymes in the absence of mass transfer effects, PITCHER [4] has summarized the integrated rate equations for plug flow and backmix reactors for the cases of Michaelis-Menten kinetics, reversible Michaelis-Menten kinetics, substrate inhibition and competitive product inhibition (Table 5.3). The table shown here is modified from PITCHER so that the definition of v is $- dS/dt$ instead of the definition used by PITCHER, namely $- dS/dt$ multiplied by the volume of the substrate solution.

Many special kinetic equations have been developed to describe particular systems. For example, TYAGI and GHOSE [24] present an equation for immobilized *Saccharomyces cerevisiae* in the conversion of cane molasses to ethanol. Their kinetic equation allows for the combined effects of inhibition by product (ethanol) and high substrate concentration.

The requirement to minimize the effect of product inhibition led DALE et al. [25, 26] to an immobilized cell reactor design in which the product is removed from the system resulting in a final, exhausted liquid phase containing ideally no sub-

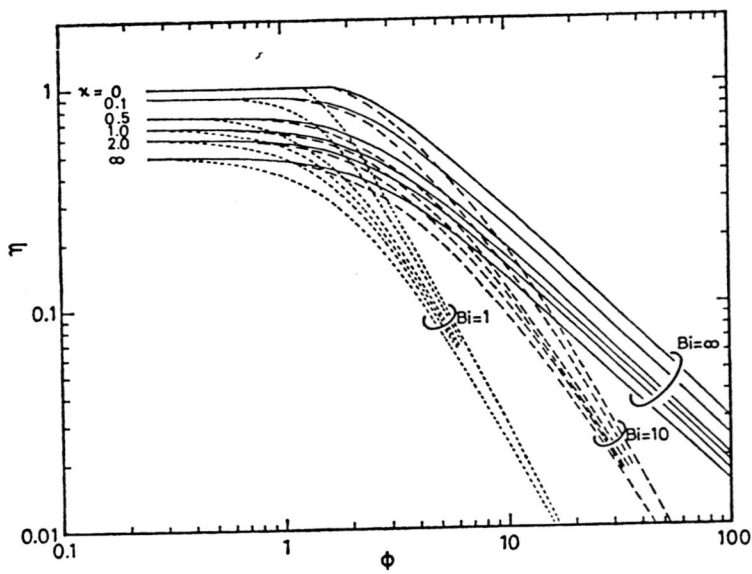

Fig. 5.6. η vs. ϕ for irreversible Michaelis-Menten kinetics when $K_p = 0.5$ and in spherical geometry. The parameter $\kappa = K_m/S_L$ [12]

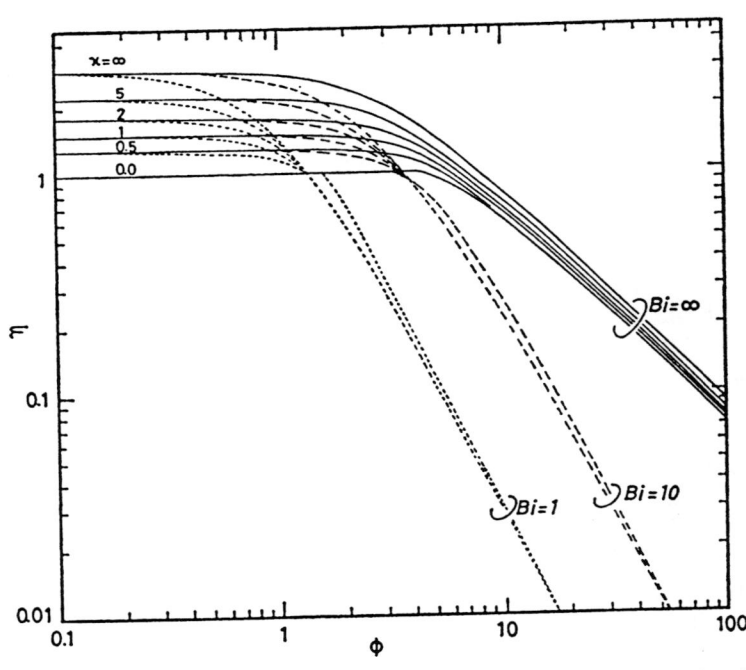

Fig. 5.7. η vs. ϕ for irreversible Michaelis-Menten kinetics when $K_p = 3.0$ and in spherical geometry. The parameter $\kappa = K_m/S_L$ [12]

144

Table 5.3. Integrated rate equations in the absence of mass transfer effects [4]

Kinetic type and rate equation

Michaelis-Menten

$$v = \frac{V_{max}S}{K_m + S}$$

Plug flow $V_{max}\tau = S_oX - K_m \ln(1-X)$

Backmix $V_{max}\tau = X\left(\dfrac{K_m}{1-X} + S_o\right)$

Reversible Michaelis-Menten

$$v = \frac{V_{max}(S - P/K)}{K_m + S - K_mP/K_r}$$

Plug flow $V_{max}\tau = X_eS_t\left\{(1 - K_m/K_r)(X_t - X_i) + \dfrac{K_m}{S_t} + 1 - X_e\dfrac{K_mK}{K_r}\ln\dfrac{X_e - X_i}{X_e - X_t}\right\}$

Backmix $V_{max}\tau = \dfrac{(X_t - X_i)(K_m + S_t - X_tS_t + K_mS_tX_t/K_r)}{1 - X - X_t/K}$

Substrate inhibition

$$v = \frac{V_{max}}{1 + K_m/S + S/K'_m}$$

Plug flow $V_{max}\tau = S_oX - K_m \ln(1-X) + \dfrac{S_0^2X}{K'_m} - \dfrac{S^2X^2}{2K'_m}$

Backmix $V_{max}\tau = XS_o\left\{1 + \dfrac{K_m}{S_o(1-X)} + \dfrac{S_o(1-X)}{K'_m}\right\}$

Competitive product inhibition

$$v = \frac{V_{max}S}{S + K_m(1 + P/K_i)}$$

Plug flow $V_{max}\tau = S_o(1 - K_m/K_i)(X_t - X_i) - \left(K_m + \dfrac{M_mS_t}{K_i}\right)\ln\left(\dfrac{1 - X_e}{1 - X_i}\right)$

Backmix $V_{max}\tau = \dfrac{(X_t - X_i)\left\{S_t(1 - X_t) + \dfrac{K_m + K_mX_tS_t}{K_i}\right\}}{1 - X_t}$

where τ is the space-time and is defined as

$$\tau = \frac{\text{reactor volume}}{\text{volumetric feed rate}}$$

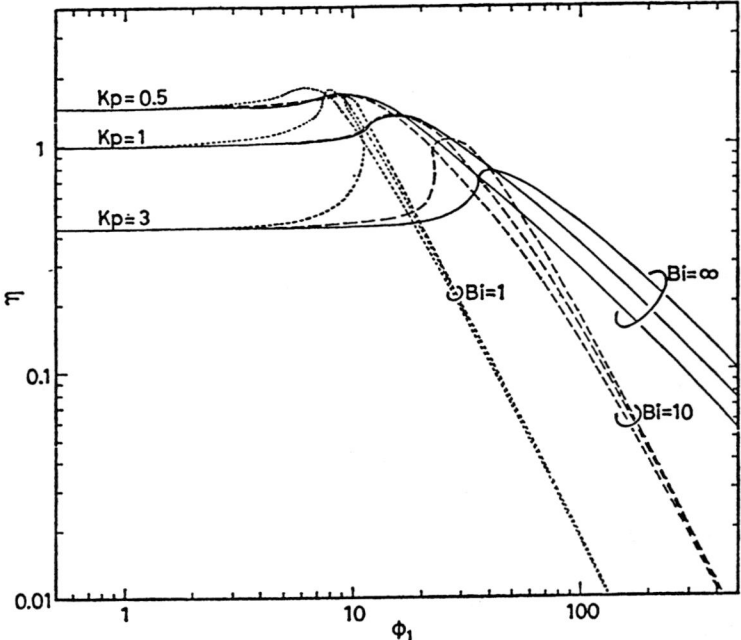

Fig. 5.8. η vs. ϕ_1 for an irreversible substrate-inhibition kinetics in spherical geometry [12]

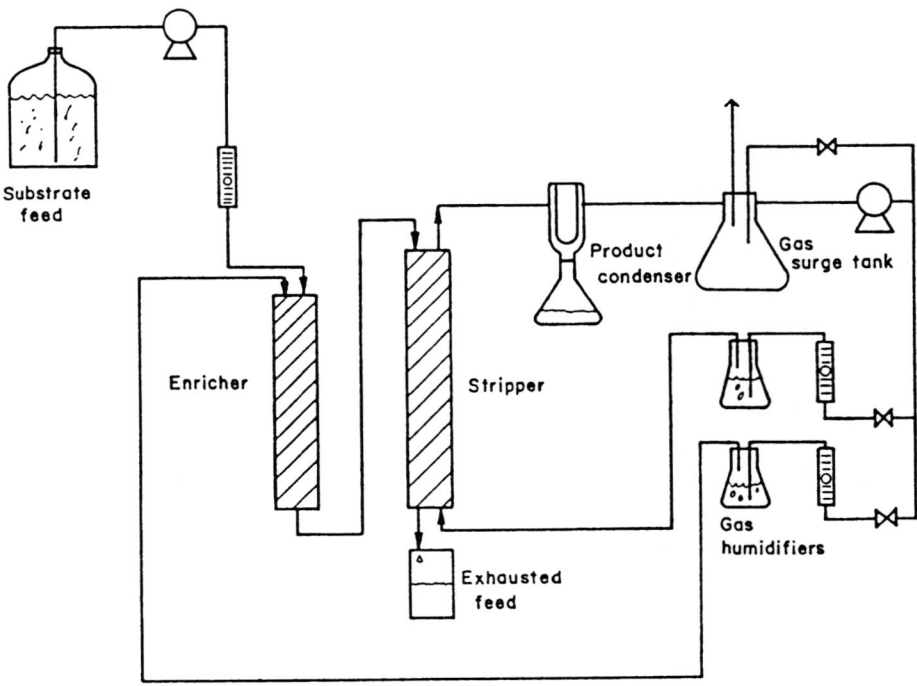

Fig. 5.9. Immobilized cell reactor-separator experimental apparatus [25]

strate or product. The reactor consists of a co-current enricher column, in which some fraction of the substrate is converted into a volatile product by the cells and moves into the gas phase according to the gas-liquid equilibrium. The liquid then enters a counter-current stripping column in which further substrate conversion occurs and the volatile product is stripped out to the gas phase. The configuration is shown in Fig. 5.9.

In addition to mass transfer from the bulk fluid to the particle surface and within the particle itself, mass transfer to the bulk phase may be important, for example, in the case of oxygen supply through a gas-liquid contactor. Mass transfer considerations for this system have been reviewed extensively [27].

Many different geometries for immobilized cell reactors have been explored. Geometries include continuously stirred tank reactors [28–32], fixed beds [33–38], fluidized beds [39–54], spouted beds [55], rotating disks [56, 57], hollow fibres [58–65], and membranes [66, 67]. Air-lift fermenters [68] do not seem to have been used for immobilized cells, but have obvious potential. The special case of choice of reactor for immobilized plant cells has been examined [69]. Which reactor is preferred depends in any given situation to a considerable extent on the physical properties of the immobilizing medium.

Notation

A	surface area of particle containing immobilized cells
Bi	$k_s R/D_e$, Biot number
d_p	particle diameter
D	diffusion coefficient
D_e	effective diffusion coefficient
k_1	intrinsic rate constant for first order reaction
k_1'	observed rate constant for first order reaction
k_s	mass transfer coefficient
K	equilibrium constant
K_i	product inhibition constant
K_m	Michaelis constant
K_m'	substrate inhibition constant
K_p	partition coefficient
K_r	constant (Michaelis type for reverse reaction)
L	plate thickness
M_m	generalized modulus based on Michaelis-Menten kinetics
M_R	generalized modulus based on observed reaction rate
P	product concentration
r	radial coordinate
R	radius of particle
S	substrate concentration
S_L	substrate concentration in the bulk fluid
S_o	initial substrate concentration
S_R	substrate concentration in the surface region of the particle
S_t	total substrate concentration if all product converted to substrate
Sh	$k_s d_p/D$, Sherwood number
t	time
v_c	rate of reaction based on a unit volume of particle containing immobilized cells
v_{free}	rate of reaction of free cells

v_{immob} rate of reaction of immobilized cells
v_r rate of reaction at radius r
V_c volume of particle containing immobilized cells
V_{max} maximum rate of reaction
X $(S_o - S)/S_o$, fractional conversion
X_e X_t at equilibrium
X_i $(S_t - S_o)/S_t$
X_t $(S_t - S)/S_t$
κ K_m/S_L in Michaelis-Menten equation (Figs. 5.6 and 5.7)
η effectiveness factor
η_{op} operational effectiveness factor
η_{ov} overall effectiveness factor, based on $S = S_L$
ϕ Thiele modulus
τ space time

References

1. Levenspiel O (1972) Chemical reaction engineering, 2nd edn. Wiley, New York
2. Smith JM (1981) Chemical engineering kinetics, 3rd edn. McGraw-Hill, New York
3. Froment GF, Bischoff KB (1979) Chemical reactor analysis and design. Wiley, New York
4. Pitcher WH Jr (1978) Adv Biochem Eng 10:1
5. Bird RB, Stewart WE, Lightfoot EE (1960) Transport phenomena. Wiley, New York
6. Furusaki S, Seki M (1985) J Chem Eng Japan 18:389
7. Seki M, Furusaki S (1985) J Chem Eng Japan 18:461
8. Monbouquette HG, Ollis DF (1986) Ann New York Acad Sci 469:230
9. Buchholz K (1982) Adv Biochem Eng 24:39
10. Fink DJ, Na T-Y, Schulz JS (1973) Biotechnol Bioeng 15:879
11. Chen K-C, Suga K, Taguchi H (1980) J Ferment Technol 58:439
12. Yamane T, Araki S, Sada E (1981) J Ferment Technol 59:367
13. Bischoff KB (1965) AIChE J 11:351
14. Goldstein L, Levin Y, Katchalski E (1964) Biochem 3:1913
15. Wharton CW, Crook EM, Brocklehurst K (1968) Eur J Biochem 6:572
16. Bunting PS, Laidler KJ (1972) Biochemistry 11:4477
17. Hinberg I, Korus R, O'Driscoll KF (1974) Biotechnol Bioeng 16:943
18. Sonomoto K, Tanaka A, Omata T, Yamane T, Fukui S (1979) Eur J Appl Microbiol Biotechnol 6:325
19. Kobayashi T, Moo-Young M (1972) Can J Chem Eng 50:162
20. Atkinson B, Davies IJ (1974) Trans Instn Chem Engrs 52:248
21. Gondo S, Isayama S, Kusunoki K (1974) J Chem Eng Japan 7:64
22. Kobayashi T, Ohmiya K, Simizu S (1976) J Ferment Technol 54:260
23. Yamane T (1981) J Ferment Technol 59:375
24. Tyagi RD, Ghose TK (1982) Biotechnol Bioeng 24:781
25. Dale MC, Okos MR, Wankat PC (1985) Biotechnol Bioeng 27:932
26. Dale MC, Okos MR, Wankat PC (1985) Biotechnol Bioeng 27:943
27. Moo-Young M, Blanch HW (1981) Adv Biochem Eng 19:1
28. Takamatsu S, Yamashita K, Sumi A (1980) J Ferment Technol 58:129
29. Furui M (1985) J Ferment Technol 63:467
30. Luong JHT (1985) Biotechnol Bioeng 27:280
31. Holladay DW, Hancher CW, Chilcote DD, Scott CD (1978) AIChE Symp Ser 74:241
32. Agrawal D, Jain VK (1986) Biotechnol Lett 8:67
33. Lortie R, Thomas D (1986) Biotechnol Bioeng 28:1256
34. Jain WK, Toran-Diaz I, Baratti J (1985) Biotechnol Bioeng 27:613
35. Sitton OC, Magruder GC, Book NL, Gaddy JL (1980) Biotechnol Bioeng Symp 10:213
36. Furusaki S, Okamura Y, Miyauchi T (1982) J Chem Eng Japan 15:148

37. Young JC, Dahab MF (1982) Biotechnol Bioeng Symp 12:303
38. Linko Y-Y, Kautola H, Uotila S, Linko P (1986) Biotechnol Lett 8:47
39. Kaletunc G, Dogu T (1983) Proc 33rd Can Soc Chem Eng Conf, Toronto, Canada, October 2–5, p 436
40. Berk D, Behie LA, Jones A, Lesser BH, Gaucher GM (1984) Proc 34th Can Soc Chem Eng Conf, Quebec City, Canada, September 30–October 3, p 279
41. Berk D, Behie LA, Jones A, Lesser BH, Gaucher GM (1984) Can J Chem Eng 62:112
42. Berk D, Behie LA, Jones A, Lesser BH, Gaucher GM (1984) Can J Chem Eng 62:120
43. Dunn IJ, Tanaka H, Uzman S, Denac M (1983) Ann New York Acad Sci 413:168
44. Bauer W (1986) Can J Chem Eng 64:561
45. Pitt WW Jr, Hancher CW, Hsu HW (1978) Am Inst Chem Engrs Symp Ser 74
46. Tanaka M, Kawaide A, Matsuno R (1986) Biotechnol Bioeng 28:1294
47. Hogrefe W, Grossenbacher H, Cook AM, Hütter R (1986) Biotechnol Bioeng 28:1577
48. Beck M, Kiesser T, Perrier M, Bauer W (1986) Can J Chem Eng 64:553
49. Dueck CL, Neufeld RJ, Chang TMS (1986) Can J Chem Eng 64:540
50. Hsu HW (1978) Biotechnol Bioeng Symp, no 8. Wiley, New York, p 1
51. Andrews GF, Przezdziecki J (1986) Biotechnol Bioeng 28:802
52. Vallat I, Monsan P, Riba JP (1986) Biotechnol Bioeng 28:151
53. Scott CD (1983) Biotechnol Bioeng Symp, no 13. Wiley, New York, p 287
54. Shieh WK, Keenan JD (1986) Adv Biochem Eng/Biotechnol 33:131
55. Webb C, Fukuda H, Atkinson B (1986) Biotechnol Bioeng 28:41
56. Del Borghi M, Converti A, Parisi F, Ferraiolo G (1985) Biotechnol Bioeng 27:761
57. Strom PF, Chung J-C (1985) Adv Biotechnol Proc 5:193
58. Kan JK, Shuler ML (1978) AIChE Symp Ser 74:31
59. Kleinstreuer C, Agarwal SS (1986) Biotechnol Bioeng 28:1233
60. Klei HE, Sundstrom DW, Coughlin RW, Ziolkowski K (1981) Biotechnol Bioeng Symp, no 11. Wiley, New York, p 593
61. Miyawaki O, Nakamura K, Yano T (1982) J Chem Eng Japan 15:224
62. Tharakan JP, Chau PC (1986) Biotechnol Bioeng 28:1064
63. Katoaka H, Saigusa T, Mukataka S, Takahashi J (1980) J Ferment Technol 58:431
64. Miyawaki O, Nakamura K, Yano T (1982) J Chem Eng Japan 15:142
65. Park TH, Kim IH, Chang HN (1985) Biotechnol Bioeng 27:1185
66. Gekas VC (1986) Enzyme Microbial Technol 8:450
67. Furusaki S, Miyauchi T (1981) J Chem Eng Japan 14:479
68. Onken U, Weiland P (1983) Adv Biotechnol Proc, vol 1. Alan R Liss, New York, p 67
69. Shuler ML, Hallsby GA, Pyne JW Jr, Cho T (1986) Ann New York Acad Sci 469:270

Author Index

152

153

154

155

157

Subject Index

160

162

Serratia marcescens 17
Serum 3, 40
–, enzyme 3
Shear stress 117
Sherwood number 134, 135, 138
Side chain degradation 27
Silica 60, 120
–, gel 45
–, hydrogel 38
Silicone 120
Slab geometry 142
Sodium alginate 106
–, phosphate 31
–, sulphite 23
Soil 4, 6, 45, 64
Solanum 80
–, *aviculare* 42, 77
Solvent 78
L-Sorbose 31, 85
L-Sorbosone 31
Space-time 145
Spherical geometry 142, 143, 144, 146
–, particle 135
Spheroplasts 82
Spinach chloroplast 15, 82
Sporasarcina urea 44
Spores 83
Spouted beds 147
Starch 26, 107
–, bouillon 28
Steady state diffusion 135
Stemphylium loti 44
Steroid 27, 35, 77, 142
–, dehydrogenation 27
–, M-ketone reduction 27
–, 9α-hydroxylation system 107
–, -Δ'-dehydrogenase 19, 27, 35
–, 1-dehydrogenase 47
–, tranformations 27
Sterols 27
Stirred tank reactors 139
Streptococcus 91
Streptococcus pyogenes 45
Streptomyces 11, 12, 13, 44, 91
–, *aureofaciens* 25
–, *clavuligerus* 28, 42, 59
–, *fradiae* 26, 106
–, *phaeochromogenes* 24, 38, 39, 56, 58
–, *tendae* 19, 107
Subcellular components 82
–, materials 81
Substrate availability 96
–, concentration 134
–, inhibition 143, 145, 146
Sucrase 3
Sucrose 14, 15, 23, 25, 30, 31, 42, 44, 82, 107

Sugar 47
–, cane 47
Sulphate 1
Sulphur 1
Surface potential 119
–, properties 118
Surfactant 39, 77
Synthetase 29
Synthetic organic carriers 123
–, polymers 26, 27, 28, 29, 30, 31, 32, 33, 34, 35, 36, 37, 38, 39, 57, 60

Talc 3
Tannin 43, 85, 120
Testosterone 33
Tetrahymena pyriformis 57
Thermal properties 117
Thiele modulus 134, 135, 137, 138, 139
Thiobacillus 1
L-Threonine 37
–, deaminase 37
Thylakoids 82
Tissue culture 4
Titanium 120
–, chloride 45
–, (IV) hydroxide 45
–, (IV)oxide 45
Toluene 78
Tortuosity 135
Toxins 3
Transformations 28
Tri-acetate beads 18
Tricalcium phosphate 16
Trichoderma 13, 107
–, *reesei* 18, 35
α,ω-Tridecanedionic acid 17
Trigonopsis variabilis 20
Trimethylammonium glycol chitosan iodide 44
Trypsin 3, 142
Tryptophan 29, 77
–, synthetase 107
L-Tryptophan 29
Tryptophanase 29
Tuber hydrolysate 47
Tuff granules 25
Tusarium 82
Tylosin 19, 107
Tyrosine 24

Uracil arabinoside 34
Urea 14, 18
Urease 3, 18
Urethane 33, 34, 36
–, prepolymers 36
Urocanic acid 43